THE COMPLETE GUIDE TO
AMAZING
PLACES

Sandy Creek
NEW YORK

An Imprint of Sterling Publishing
1166 Avenue of The Americas
New York, NY, 10036

ISBN 978-1-4351-6159-7

Manufactured in China

Lot #:
2 4 6 8 10 9 7 5 3

11/15

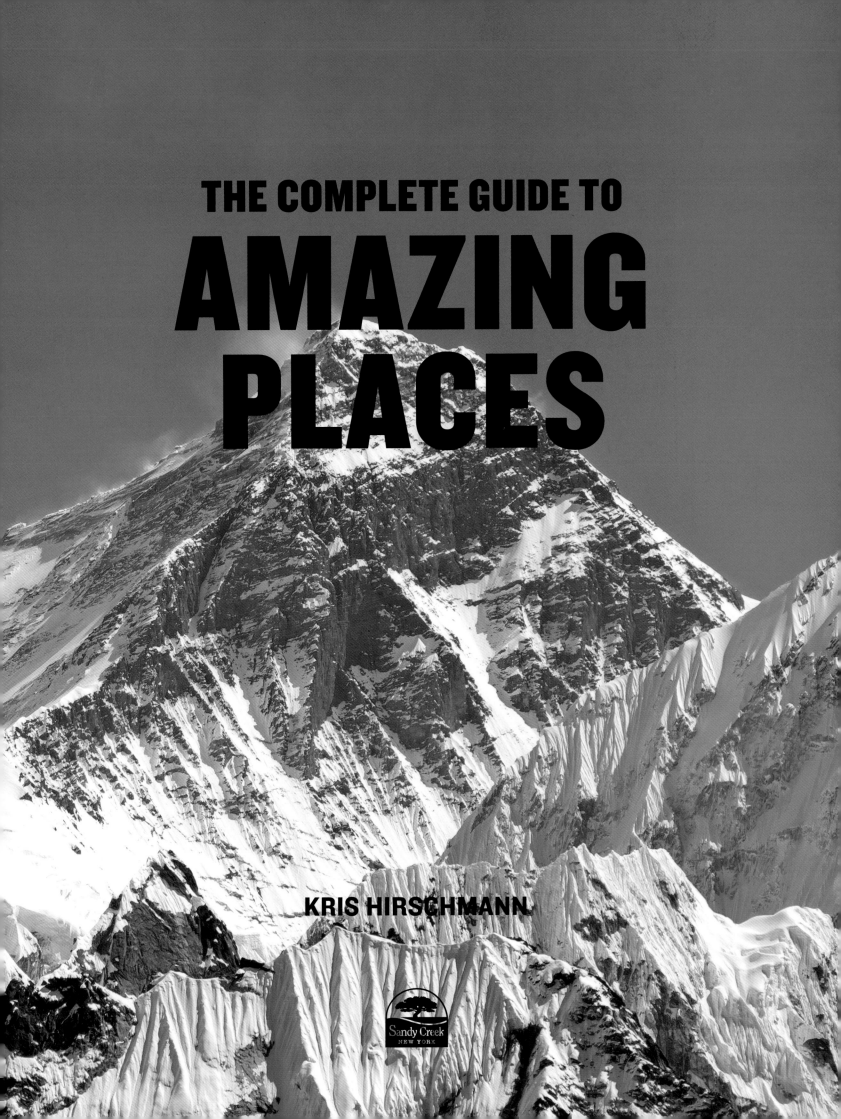

THE COMPLETE GUIDE TO
AMAZING
PLACES

KRIS HIRSCHMANN

Sandy Creek
NEW YORK

CONTENTS

Words in **bold** are explained in the Glossary on page 138.

ANTARCTICA

Antarctica is a massive, ice-covered continent. It is the world's most southerly place. With an average altitude of 7,500 feet, it is, on average, the world's highest and coldest continent. In the 19th century, several expeditions penetrated this remote, icy world. Today, 30 countries maintain permanent or summer-only research stations in Antarctica.

Glacier

Gentoo penguins enjoy a ride on an iceberg in Antarctica.

Glaciers

Antarctica is home to many **glaciers**. A glacier is a huge, slow-moving mass of ice. Glaciers can have deep cracks or **crevasses** in them—sometimes as deep as the glacier itself. These cracks are dangerous for mountaineers, scientists, and explorers.

6

Ice Sheet

Lying over most of Antarctica is the biggest mass of ice in the world. The Antarctic ice sheet covers an area as big as the United States and Europe put together. It is more than 2.5 miles deep in some places, and it contains 70 percent of the world's fresh water.

AMAZING!

In 1983, a research station in Antarctica recorded the coldest temperature ever measured on Earth: -128.6°F!

Only around 12 percent of an iceberg is above the surface.

Icebergs

An **iceberg** is a huge, floating chunk of ice. Icebergs form when chunks of ice break off from a glacier, ice sheet, or large iceberg. Ice is slightly lighter than water. This means that icebergs float, but just barely.

MENDENHALL GLACIER

Mendenhall Glacier is a massive river of ice near the Alaskan city of Juneau. This frozen feature is about 12 miles long, 1.5 miles wide, and 100 feet deep. It flows from its source in the Juneau Icefield to Mendenhall Lake, advancing several feet per day. Chunks of the glacier regularly crack off and tumble into the lake in a process called **calving**.

The blue ice of the Mendenhall Glacier.

Ice Caves

Mendenhall Glacier is famous for its spectacular ice caves. The caves were carved by water flowing through the glacier. To reach the caves, visitors must kayak across a stretch of near-freezing lake and then make a dangerous hike across the glacier's cracked surface. Inside the caves, falling ice is a constant hazard.

Reaching the caves is a challenge!

Blue Ice

Like all glaciers, Mendenhall Glacier is blue because its ice is highly compressed. The glacier's weight pushes air bubbles out of the ice and makes its crystals bigger. The dense ice absorbs all colors of the **visible spectrum** except blue. This color is reflected, making the ice look blue to observers.

AMAZING!

Some ancient trees stand inside the Mendenhall ice caves. The glacier is revealing frozen forests as it melts!

The face of the glacier. Climate change is causing glaciers to melt faster than usual

MOUNT EVEREST

Mount Everest, the world's highest mountain, rises 29,035 feet above sea level and is a massive snow- and ice-topped chunk of rock in the Himalayas, a huge mountain range in Asia. The international border between China and Nepal runs across the peak's precise **summit** point.

Conquering Everest

In 1953, Sir Edmund Hillary and Tenzing Norgay became the first climbers to reach the top of Everest and make it back home. Since then, more than 4,000 people have conquered the mountain, including a 73-year-old woman.

Sir Edmund Hillary and Tenzing Norgay after returning from the summit of Everest in 1953.

Thin Air

The amount of oxygen in the air decreases with increasing altitude. Above about 26,000 feet, the air on Mount Everest is so thin that climbers have extreme difficulty breathing. Most climbers use oxygen masks and tanks in this "death zone," as it is called.

Two Routes

There are two main routes up Mount Everest: the southeast ridge from Nepal and the north ridge from Tibet. The southeast ridge is the easier and more popular of the two.

The jagged edges jutting into the air on Everest's slopes are incredibly dangerous.

Trekkers admire the view of Everest from Gokyo Lakes in Nepal.

TORNADO ALLEY

A tornado spirals out from the clouds toward the ground.

The world's most destructive tornadoes occur in a part of the central United States known as Tornado Alley. This area stretches across South Dakota, Nebraska, Kansas, Oklahoma, and northern Texas, where cold air from Canada merges with warm, moist air from the Gulf of Mexico to create frequent twisters. Every year, tornadoes in this area destroy hundreds of homes and kill dozens of people.

Tornado Alley has all of the weather conditions needed for a tornado to take shape.

Montana · North Dakota · Minnesota · Wisconsin · Idaho · Wyoming · South Dakota · Iowa · IN · NV · Nebraska · Illinois · Utah · Colorado · Kansas · Missouri · KY · Arizona · Oklahoma · TN · New Mexico · Arkansas · AL · Texas · Louisiana

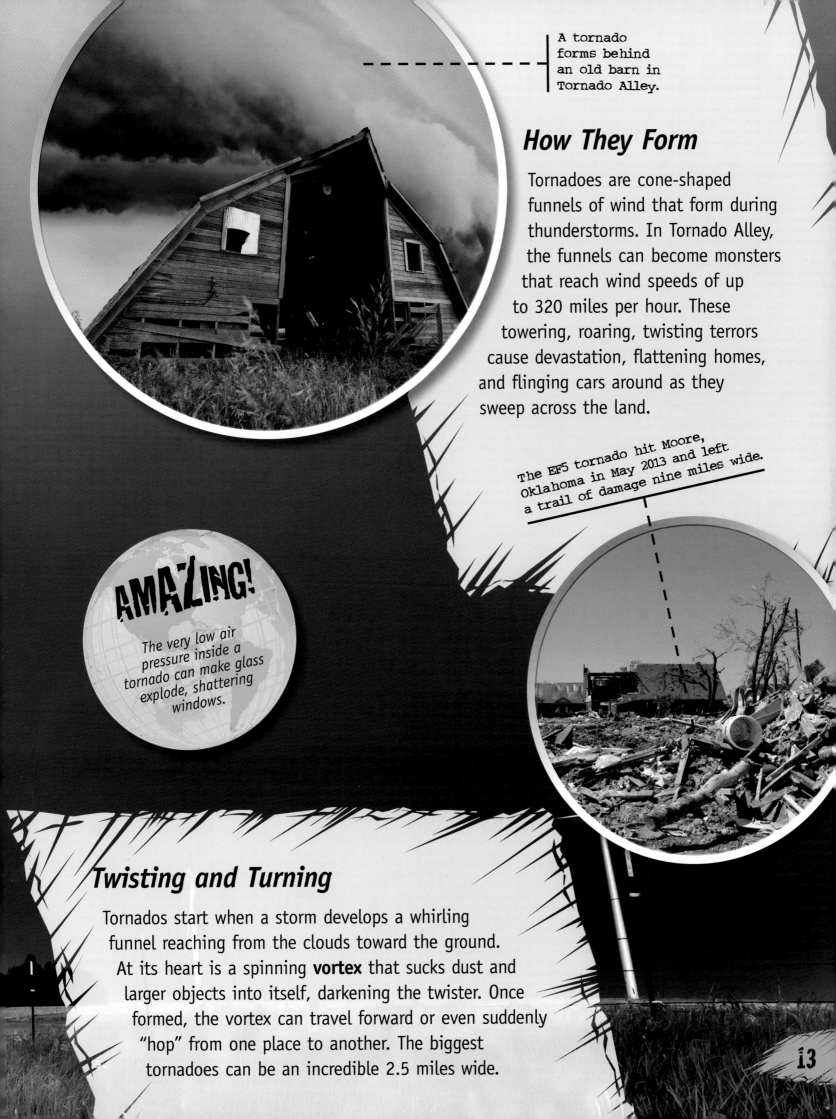

A tornado forms behind an old barn in Tornado Alley.

How They Form

Tornadoes are cone-shaped funnels of wind that form during thunderstorms. In Tornado Alley, the funnels can become monsters that reach wind speeds of up to 320 miles per hour. These towering, roaring, twisting terrors cause devastation, flattening homes, and flinging cars around as they sweep across the land.

The EF5 tornado hit Moore, Oklahoma in May 2013 and left a trail of damage nine miles wide.

AMAZING!

The very low air pressure inside a tornado can make glass explode, shattering windows.

Twisting and Turning

Tornados start when a storm develops a whirling funnel reaching from the clouds toward the ground. At its heart is a spinning **vortex** that sucks dust and larger objects into itself, darkening the twister. Once formed, the vortex can travel forward or even suddenly "hop" from one place to another. The biggest tornadoes can be an incredible 2.5 miles wide.

ANGEL FALLS

With an uninterrupted plunge of 2,648 feet and nearly 600 feet of additional rocky falls at the bottom, Angel Falls in Venezuela is the world's tallest waterfall. In total, this delicate and beautiful waterfall is 3,212 feet high—taller than the world's highest skyscraper.

Cliff Top

Angel Falls flows over an amazingly tall cliff on one side of a **tepui**, or table-top mountain—a typical feature of the Venezuelan landscape. From the top, the water drops straight down in a shimmering, misty column before reaching an area of sloped cascades and rapids.

Angel Falls in the morning light.

About 62,000 cubic feet of water gushes over Angel Falls every second.

AMAZING!

Angel Falls is named after pilot James Angel, who flew around the Falls in 1935 and first reported them to the outside world.

Water Source

Angel Falls is fed by rain and clouds. Conditions are constantly damp and wet at the top of the tepui. Moisture collects and flows off the tepui in a great stream. The volume of water in Angel Falls varies, being greatest during the rainy season (June through November) and least during the dry season (December through March).

NIAGARA FALLS

Niagara Falls is actually a group of three waterfalls on the Niagara River straddling the Canada/New York border. About 90 percent of the water plummets over Horseshoe Falls, the largest of the three. The rest flows over the much smaller American Falls and Bridal Veil Falls. Together, these drops have the highest flow rate of any waterfall or waterfall system in the world—up to 6 million cubic feet per minute.

Dug by Water

Niagara Falls formed at the end of the last ice age, around 10,000 years ago, when glaciers melted to form the U.S. Great Lakes. The lakes fed the Niagara River as excess water flowed toward the sea. Over time, the river dug a **gorge**. This gorge became the basis for the mighty waterfalls that exist today.

Horseshoe Falls in Ontario, Canada, is the biggest of the three waterfalls.

Niagara Falls from the American side.

American
Falls

Bridal Veil
Falls

Horseshoe
Falls

Cliff Erosion

The waters of Niagara Falls are **eroding** the cliff they tumble from at a rate of roughly one foot per year. This means the falls are actually moving upriver little by little. In about 50,000 years, the falls will have receded all the way to Lake Erie, and Niagara Falls will cease to exist.

AMAZING!

Since 1901, 15 people have deliberately gone over Niagara Falls in a barrel or another device. Of these daredevils, ten survived and five died.

AMAZON RIVER

The Amazon River in South America is the largest river in the world by volume. It contains as much water as the next seven biggest rivers combined! It drains an area called the Amazon Basin that covers more than 2.7 million square miles, or about 40 percent of the South American continent. That's enough to account for about one-fifth of the world's river water.

A stunning view of the curving Amazon River and the forest around it.

AMAZING!

The Amazon River is mainly surrounded by rainforest, so there are few roads around it and not a single bridge across it.

Water World

Where it flows into the Atlantic Ocean, the two banks of the Amazon River are 120 miles apart. A speedboat traveling across the river at 60 miles per hour would take two hours to reach the other side! The river stretches more than 4,000 miles, from deep in the Amazon rainforest all the way to South America's east coast.

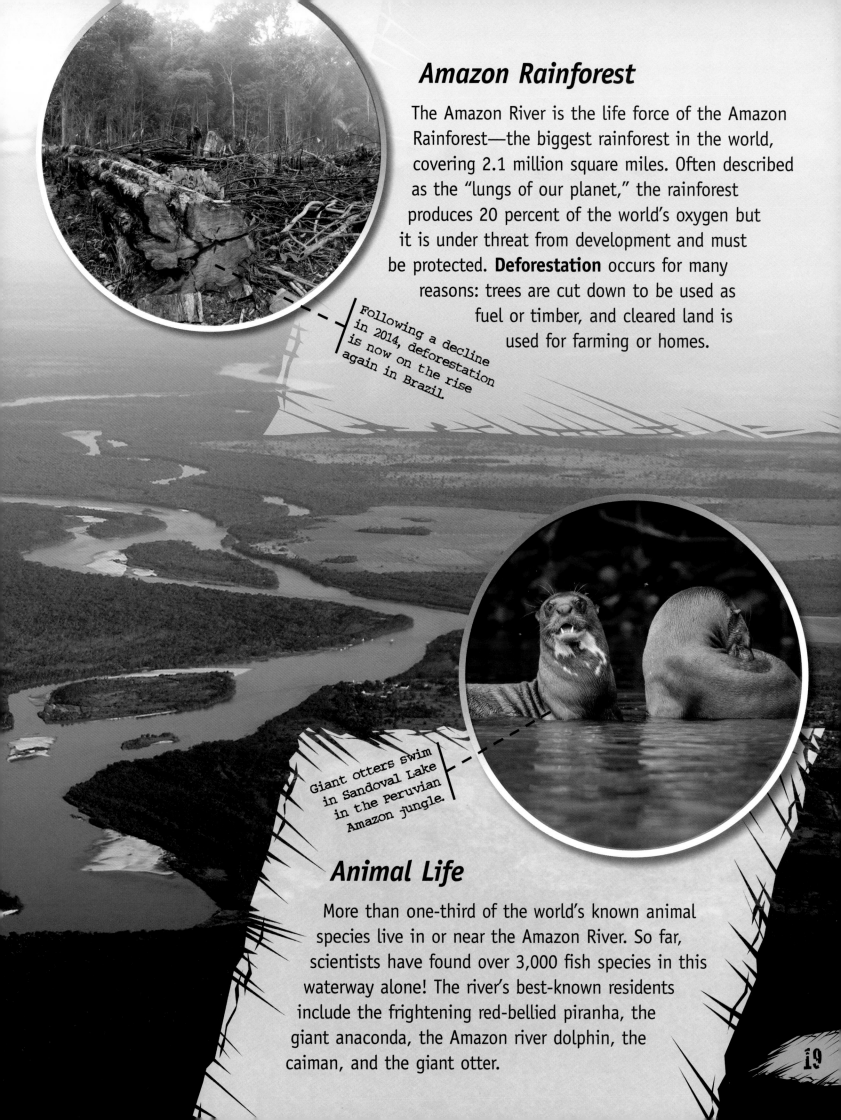

Amazon Rainforest

The Amazon River is the life force of the Amazon Rainforest—the biggest rainforest in the world, covering 2.1 million square miles. Often described as the "lungs of our planet," the rainforest produces 20 percent of the world's oxygen but it is under threat from development and must be protected. **Deforestation** occurs for many reasons: trees are cut down to be used as fuel or timber, and cleared land is used for farming or homes.

Following a decline in 2014, deforestation is now on the rise again in Brazil.

Giant otters swim in Sandoval Lake in the Peruvian Amazon jungle.

Animal Life

More than one-third of the world's known animal species live in or near the Amazon River. So far, scientists have found over 3,000 fish species in this waterway alone! The river's best-known residents include the frightening red-bellied piranha, the giant anaconda, the Amazon river dolphin, the caiman, and the giant otter.

19

NILE RIVER

Flowing northward across an incredible 4,160 miles of the African continent, the Nile is the world's longest river. This waterway begins in the Great Lakes region of central Africa and travels across 11 countries—Tanzania, Uganda, Rwanda, Burundi, Congo-Kinshasa, Kenya, Ethiopia, Eritrea, South Sudan, Sudan, and Egypt—before emptying itself into the Mediterranean Sea.

View of the Nile River from the Aswan Dam.

Cradle of Civilization

Along certain stretches, the Nile floods its banks each year. The annual floods create fertile land where farms, animals, and people can thrive. In long-ago times, this geographic quirk attracted large populations and allowed mighty civilizations to develop. The ancient Egyptian culture we know so well today could not have existed without the Nile's help.

The Ancient Egyptian temple of Luxor was built on the east bank of the Nile. This sphinx statue forms part of Luxor's Sphinx Alley.

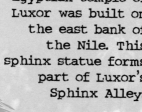

20

Flood Control

Although the Nile's floods are good for the land, they are not always good for farmers. Unusually high floods can destroy entire crops. Unusually low floods can bring droughts. To control the waters, the Aswan Dam was constructed between 1960 and 1970 on the northern border between Egypt and Sudan. The dam maintains the Nile's flow which keeps the farmers happy.

The Aswan Dam is 14 miles wide in places.

GREAT BARRIER REEF

The Great Barrier Reef is the world's biggest coral reef. It is HUGE. The reef stretches for more than 1,250 miles along the northeast coast of Australia, forming a chain of coral islands and underwater platforms that can be seen from space. The system is one of the world's most diverse **ecosystems** and one of the seven natural wonders of the world.

Underwater Wonderland

While some coral reefs become islands, others are still underwater, where they provide a **habitat**, or home, for all kinds of sea creatures. Divers may see turtles, octopuses, sea snakes, jellyfish, dolphins, and more than 1,000 species of fish, including more than 100 types of sharks and rays.

An aerial photograph of the Great Barrier Reef.

The Great Barrier Reef has between 1,500 and 2,000 species of fish.

Coral Structure

Coral is a kind of skeleton made by tiny sea creatures called coral **polyps**, which are sea anemones. Each polyp builds a skeleton around itself using sea **minerals**. Each generation builds more coral on top of the old coral, and gradually a big structure, or reef, builds up. There are many different species of coral polyps.

A scuba diver observes some of the coral of the Great Barrier Reef.

AMAZING!

The Great Barrier Reef is the largest construction on the planet to be built by living things.

THE EVERGLADES

The Everglades is a tropical **wetland** region in southern Florida in the United States. Covering about 1.5 million acres, this massive area contains many ecosystems, including freshwater and saltwater habitats, prairies, hardwood forests, mangrove forests, and even coral reefs. It is one of America's wildest and hardest-to-navigate places.

Many slender waterways wind through sawgrass and other vegetation in the Everglades.

A cypress forest in the Everglades, one of the many ecosystems in the wetland region.

River of Grass

Lake Okeechobee, in the center of Florida, is the Everglades' main source of water. During the wet season, the lake overflows. The water runs southward to form a slow-moving, shallow river up to 60 miles wide. Sawgrass pokes up through the river, giving this waterway the nickname "River of Grass."

A marsh area of Lake Okeechobee in the Central Everglades.

AMAZING!

The Everglades are home to 27 different types of snakes, including the massive Burmese python, which can grow up to 22 feet long.

Burning fields in the Florida Everglades.

The Role of Fire

Fire plays an important role in the Everglades. Frequent lightning strikes in this area cause forest fires that burn away dead branches, leaves, and other plant material without harming fire-resistant plants. Fire's heat also "wakes up" certain seeds and pinecones and starts their growing process. These effects and others keep the Everglades healthy.

LAKE BAIKAL

AMAZING!

336 rivers flow into Lake Baikal, but only one—the Angara River—flows out.

Nestled in a mountainous region of southern Siberia, Lake Baikal is amazing in many ways. It is the world's deepest lake, reaching a maximum depth of 5,387 feet. It is the largest lake by volume, containing about 20 percent of Earth's unfrozen surface freshwater. It is also thought to be our planet's oldest lake, with an estimated age of 25 to 30 million years.

Every winter Lake Baikal freezes over and people can ride horses over it safely.

Baikal Seals bathing in the sun.

Weird Wildlife

Living in Lake Baikal are more than 1,000 species of wildlife that are not found anywhere else in the world. One of these is the Baikal seal, or nerpa. The nerpa feeds on a strange, transparent fish, called the golomyanka, which is also found only in Lake Baikal.

Infant Ocean

Lake Baikal is located in a rift valley, which is a place where two of Earth's continental plates are pulling apart. Today, the lake is spreading at a speed of about 1 inch per year. Millions of years from now, Lake Baikal will be the Baikal Ocean!

Olkhon Island is the largest island on Lake Baikal.

THE GREAT BLUE HOLE

The Great Blue Hole, off the coast of Belize in Central America, measures over 1,000 feet across and 412 feet deep. This awesome natural formation is a huge, deep hole in the seabed. It appears dark blue because it is so much deeper than the surrounding shallow water.

There are other blue holes, but the Great Blue Hole is the biggest and most impressive.

From Cave to Hole

Experts think the Great Blue Hole formed around 10,000 years ago. The hole was then an enormous underground cave, covered by a rocky roof. When the last ice age ended and sea levels rose, the roof collapsed and seawater filled the hole. There are **stalactites** and other amazing rock formations on the hole's walls, which probably formed when it was still a dry cave.

Sinkholes

The Great Blue Hole is an underwater **sinkhole**. Sinkholes can open up on land, too—and sometimes they appear without any warning. If this happens in a busy, built-up area, it can be disastrous. When this sinkhole opened up in Guatemala City, it swallowed 12 houses and killed three people.

This sinkhole appeared in Guatemala City, Guatemala, in 2007.

29

DEEP-SEA VENTS

Miles below the ocean's surface, the water is pitch-black and ice cold. Scientists knew, though, that a few deep-sea areas contained extra-warm water. In 1977, they went searching for an explanation for this puzzling fact—and what they found was amazing!

Amazing Discovery

Exploration vessels found groups of chimney-like structures—which came to be called **hydrothermal vents**—pumping dark, hot water into the sea. Around the strange structures were life forms no one had ever seen before, from tiny white crabs and giant mussels to worms and snails.

AMAZING!

The biggest known black smoker is 160 feet high. Scientists have nicknamed it Godzilla.

A deep-sea diving vessel called Alvin was used to explore the sea floor.

Black water shoots out from a hydrothermal vent.

Black Smokers

Hydrothermal vents are also called black smokers. These underwater chimneys belch out hot water containing minerals from the ocean floor. These minerals support communities of animals whose bodies can change chemicals into food. This process is called **chemosynthesis**.

Tube worms at a hydrothermal vent site.

Tube Worms

These weird red-gilled white worms can grow up to ten feet long! They have special **bacteria** in their bodies that trap chemicals from the hot vent water.

31

THE DEAD SEA

The Dead Sea, located between Israel and Jordan, is actually not a sea but a huge salt lake. It lies 1,300 feet below sea level. This makes the Dead Sea different from most lakes, which lose water through rivers flowing out to the ocean. Since rivers cannot flow uphill, the Dead Sea keeps its water. The liquid **evaporates** in the region's hot sunshine, leaving its salt and minerals behind.

AMAZING!

Many people believe that the salt and minerals in the Dead Sea can help to alleviate skin diseases and other conditions.

The Dead Sea is also known as the Salt Sea (despite being a huge lake).

Extra Salt

The Dead Sea is more than 34 percent salt, compared to about 3 percent for normal seawater. The salt content makes the water very dense. People float on it like beach balls floating on a pool! Pebbles of halite, a natural salt, line the beaches along the water's edge.

The Dead Sea's dense water makes swimmers float high up.

No Life

Although a few bacteria and fungi manage to live in the Dead Sea, larger plants and animals cannot survive in the lake's salty waters. The water's "dead" appearance gives the lake its name. The mountains surrounding the Dead Sea, however, teem with life. Visitors might see camels, foxes, leopards, ibex, jackals, or hundreds of different bird species.

BONNEVILLE SALT FLATS

Bonneville Salt
Flats at sunset.

A salt flat is a completely flat stretch of hard, solid salt. Visitors can experience this natural marvel first-hand at Utah's Bonneville Salt Flats. This salt plain is one of the world's largest, covering an area of about 47 square miles. On sunny days, the ground's thick, salty crust glistens like snow as far as the eye can see.

AMAZING!

Every winter, a shallow layer of water floods the Bonneville Salt Flats. When warm weather arrives, the water evaporates and a new, perfectly flat surface appears.

The Bonneville Salt Flats are flooded by winter rains every year.

What's a Salt Flat?

Salt flats form when salt and other minerals collect in a lake and the water then evaporates. The Bonneville Salt Flats are the remains of ancient Lake Bonneville, which dried up over 14,000 years ago.

The World of Speed racing car event takes place at Bonneville every year.

The Bonneville Speedway

The Bonneville Speedway is located on an especially smooth part of the Bonneville Salt Flats. The Speedway hosts several motor racing events each year. It is also the best place in the world to set land speed records. Since 1949, drivers have broken the 300-400-500-and 600-mile-per-hour land speed records at this site.

WHITE SANDS NATIONAL MONUMENT

White Sands National Monument in New Mexico is home to the world's largest gypsum **dune** field. The sparkling white dunes cover an area of about 275 square miles and rise up to 60 feet high. Blown by the area's relentless wind, the dunes travel up to 30 feet per year. They swallow up any plants in their way, turning the desert into a barren but beautiful wasteland.

Impressive gypsum dunes at the National Monument.

Plants survive only in sheltered areas, where they will not be covered by traveling dunes.

Gypsum vs. Quartz

Most sand is made from grains of quartz, which is hard and often tan in color. Gypsum crystals, on the other hand, are white and soft. They dissolve easily when moisture is present. Gypsum dunes therefore form only in the driest places.

A side-blotched lizard in the sparkling white sands.

White Animals

Many animals live in and on the gypsum dunes. Some of these animals have **evolved** over time to be pale or white to match their home environment. Lighter-than-usual insects, spiders, scorpions, lizards, mammals, and even toads can all be seen at White Sands National Monument.

ATACAMA DESERT

Chile's Atacama Desert is the driest place in the world. Average rainfall across this coastal desert is about 1/2 inch per year, but rainwater has never reached some parts of this region. Scientists believe that these conditions have existed for about three million years. That makes the Atacama one of the oldest deserts on Earth.

Rock formations in the Atacama Desert.

AMAZING!

Because of its unearthly landscape, the Atacama Desert sometimes stands in for the planet Mars in science-fiction movies.

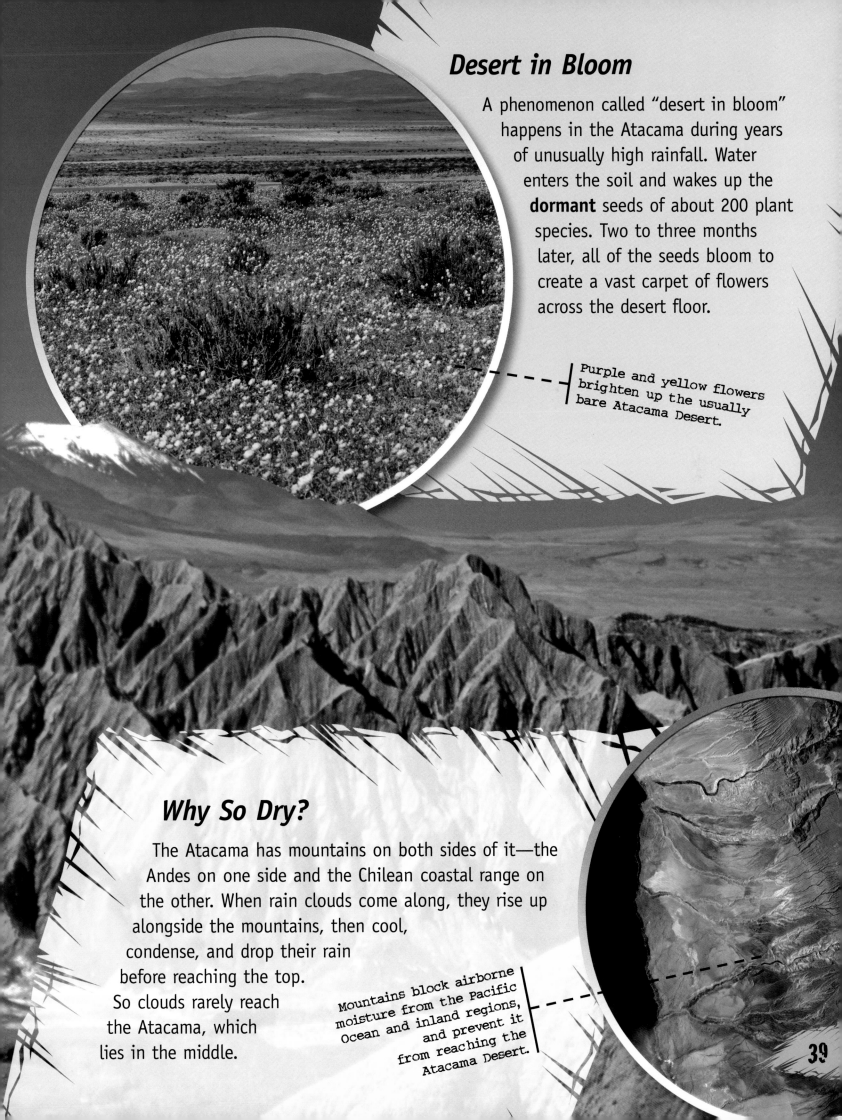

Desert in Bloom

A phenomenon called "desert in bloom" happens in the Atacama during years of unusually high rainfall. Water enters the soil and wakes up the **dormant** seeds of about 200 plant species. Two to three months later, all of the seeds bloom to create a vast carpet of flowers across the desert floor.

Purple and yellow flowers brighten up the usually bare Atacama Desert.

Why So Dry?

The Atacama has mountains on both sides of it—the Andes on one side and the Chilean coastal range on the other. When rain clouds come along, they rise up alongside the mountains, then cool, condense, and drop their rain before reaching the top. So clouds rarely reach the Atacama, which lies in the middle.

Mountains block airborne moisture from the Pacific Ocean and inland regions, and prevent it from reaching the Atacama Desert.

NAMIB DESERT

The Namib in Africa is thought to be the oldest desert on the planet—it has been as dry as a bone for around 55 million years. It is also one of the world's unique and most beautiful places, with its soaring dunes and the spooky Skeleton Coast. The pointy, red-brown dunes form when the sand shifts and rolls down one side of the dune, leaving a razorlike ridge along the top.

The stunning red dunes of Sossusvlei, in the Namib desert.

AMAZING!

The Namib Desert's biggest sand dunes are up to 980 feet high and 20 miles long!

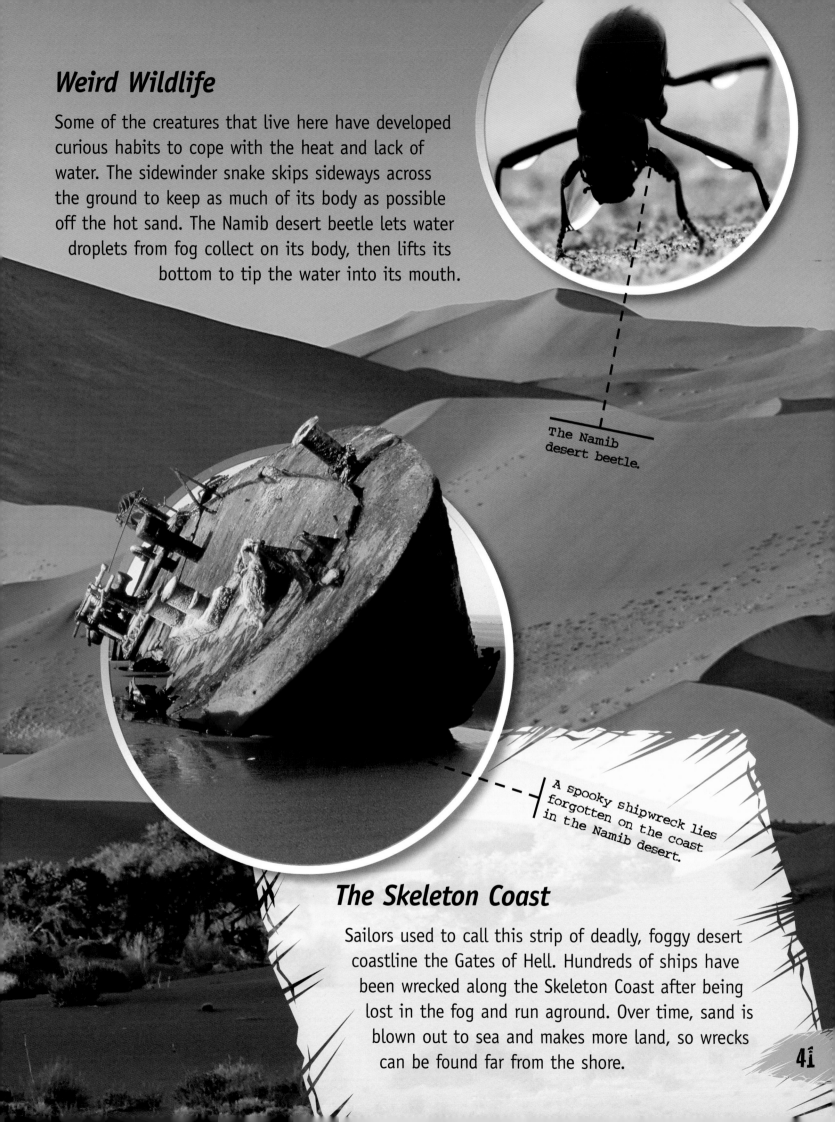

Weird Wildlife

Some of the creatures that live here have developed curious habits to cope with the heat and lack of water. The sidewinder snake skips sideways across the ground to keep as much of its body as possible off the hot sand. The Namib desert beetle lets water droplets from fog collect on its body, then lifts its bottom to tip the water into its mouth.

The Namib desert beetle.

A spooky shipwreck lies forgotten on the coast in the Namib desert.

The Skeleton Coast

Sailors used to call this strip of deadly, foggy desert coastline the Gates of Hell. Hundreds of ships have been wrecked along the Skeleton Coast after being lost in the fog and run aground. Over time, sand is blown out to sea and makes more land, so wrecks can be found far from the shore.

41

DEATH VALLEY

California's Death Valley is a place of extremes.
This desert contains Badwater Basin, the lowest point
in the United States, with an elevation of 282 feet
below sea level. It is America's driest place, with an
average rainfall of less than two inches per year.
It is also the hottest spot on Earth, having reached
a world-record 134°F on July 10, 1913.

Broadwater Basin
as seen from
Dante's View,
a viewpoint.

Why So Hot?

Death Valley is shaped like a basin surrounded by mountain ranges. During summer months, sunlight heats the basin's floor to incredible temperatures. The heat rises but cannot escape because of the surrounding mountain walls. It circulates within the valley in the form of blistering hot winds.

Death Valley National Park.

A sailing stone has left a track in the desert.

Sailing Stones

Death Valley's "sailing stones" are boulders that travel across the desert floor, leaving straight tracks behind them. They are found in a flat area called Racetrack Playa. These stones were a mystery for many decades, but scientists have recently learned that winter ice shoves them from place to place.

HAWAIIAN ISLANDS

You can't reach Kalaupapa by car, only by boat, plane, mule, or on foot.

The Hawaiian Islands is a group of eight major islands and many smaller structures in the North Pacific Ocean. The chain stretches 1,500 miles from the island of Hawai'i in the south to Kure Atoll in the north. All of the land features in the Hawaiian Islands were formed by volcanic eruptions under the sea. Some of the chain's volcanoes are still erupting today.

Kalaupapa Cliffs

On the Hawaiian island of Molokai, towering over the tiny village of Kalaupapa, are some of the world's most enormous sea cliffs. Covered with tropical plants, the cliffs plunge just over 3,280 feet down to the sea. These enormous cliffs were originally part of a tall volcano. When part of the volcano fell into the sea, the steep cliffs were left behind.

Mauna Loa

Mauna Loa is an active volcano on the island of Hawai'i. This impressive peak rises 13,678 feet above sea level, but that's not even half of the volcano's true height. Mauna Loa also stretches miles below the water. In total, this mighty mountain is 30,085 feet from base to summit, making it the world's tallest (although not the highest elevation) peak.

Mauna Loa last erupted in 1984.

KILAUEA

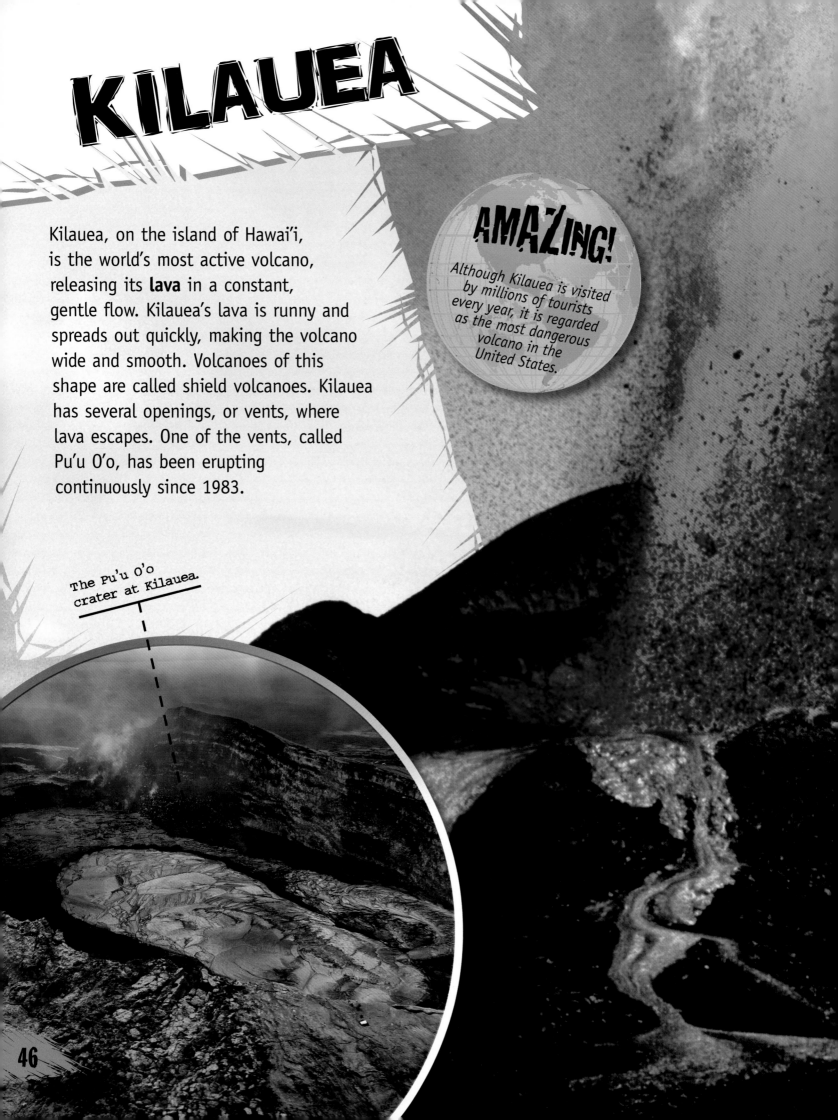

Kilauea, on the island of Hawai'i, is the world's most active volcano, releasing its **lava** in a constant, gentle flow. Kilauea's lava is runny and spreads out quickly, making the volcano wide and smooth. Volcanoes of this shape are called shield volcanoes. Kilauea has several openings, or vents, where lava escapes. One of the vents, called Pu'u O'o, has been erupting continuously since 1983.

AMAZING!

Although Kilauea is visited by millions of tourists every year, it is regarded as the most dangerous volcano in the United States.

The Pu'u O'o crater at Kilauea.

What is Lava?

Kilauea's lava starts life inside Earth as molten rock called **magma**, but when it erupts from the ground it becomes a gushing open-air stream. Lava can be up to 2,200°F—five times hotter than the hottest kitchen oven. Its heat is so intense that lava can set plants on fire as it flows past them, or boil water when it hits the sea.

Kazumura Cave

More than 40 miles long, Kazumura Cave on the eastern slope of Kilauea is the world's longest lava tube. It formed about 500 years ago when a river of lava hardened on the outside but continued to flow on the inside. The rock "tube" emptied out like a drinking straw, leaving a long cavern behind.

The world's largest lava tube at Kilauea volcano.

The Kilauea volcano erupts. Rivers of lava head to the ocean.

47

YELLOWSTONE NATIONAL PARK

Located almost entirely in the state of Wyoming, Yellowstone National Park was created in 1872. It is the world's first national park. Sitting on top of a volcanically active area, the park is famous for its hot springs, squirting geysers, and **fumaroles**, where hot gases escape from the ground. Just under the surface is a vast chamber of magma and gas waiting to erupt.

West Thumb Geyser Basin, in Yellowstone National Park.

Satellite image of the caldera.

Underground Volcano

Yellowstone's magma lies beneath a vast depression called the Yellowstone Caldera. A **caldera** is a wide, flat, bowl-shaped crater left behind after a volcanic eruption. The Yellowstone Caldera formed about 630,000 years ago—the last time the region's volcano blew. The eruption left a crater 28 miles across and 47 miles wide.

Giant Fountain

Old Faithful Geyser is one of the most famous volcanic features at Yellowstone. It shoots a jet of scalding hot water 170 feet into the air about once every 90 minutes. The water is heated by magma under the ground. It is forced to erupt as steam forms and pressure builds.

AMAZING!

Yellowstone National Park is a popular tourist destination. It attracts about two million visitors each year.

Boiling hot water shoots out from Old Faithful

49

ULURU

Uluru, also known as Ayers Rock, is a massive boulder that juts out of a flat, dry stretch of the Australian Outback. This formation is more than two miles long and one mile wide, and it rises 1,141 feet above the ground. Like an iceberg, Uluru is even bigger than it looks; the rock's base extends many thousands of feet below the earth's surface.

Uluru is in Northern Territory, Australia's Outback.

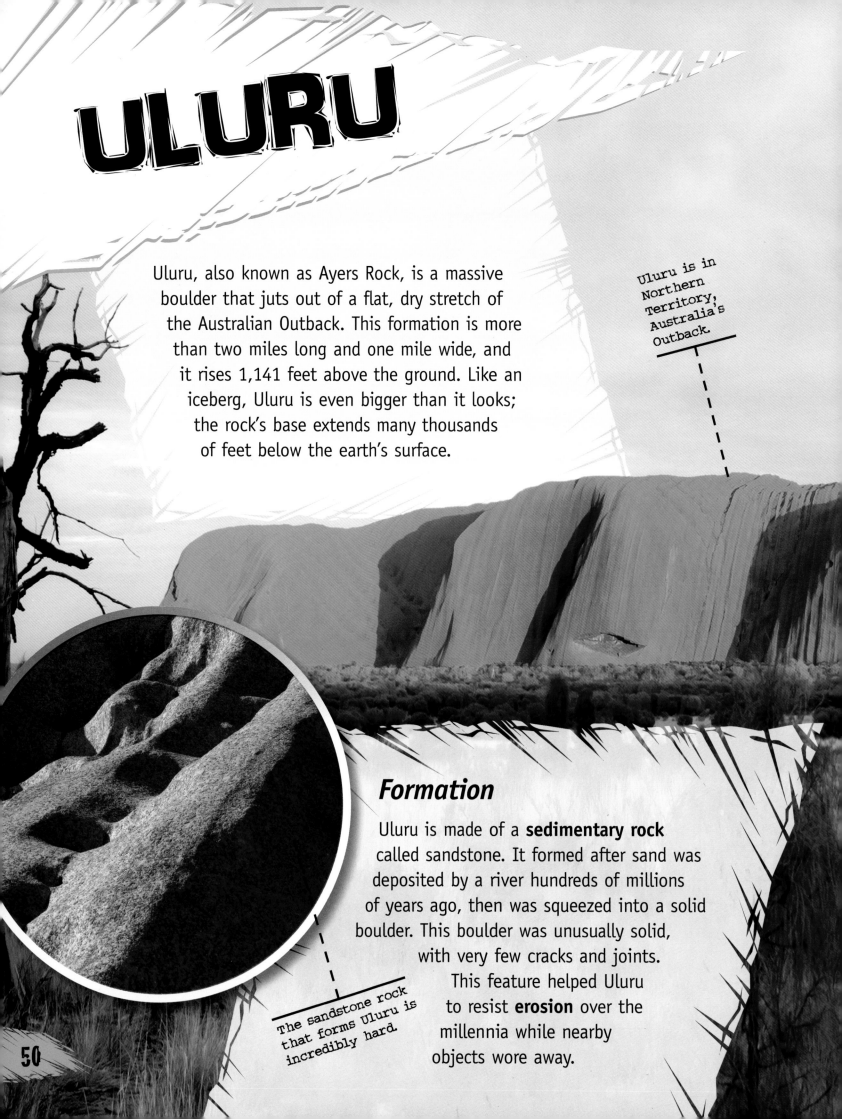

The sandstone rock that forms Uluru is incredibly hard.

Formation

Uluru is made of a **sedimentary rock** called sandstone. It formed after sand was deposited by a river hundreds of millions of years ago, then was squeezed into a solid boulder. This boulder was unusually solid, with very few cracks and joints. This feature helped Uluru to resist **erosion** over the millennia while nearby objects wore away.

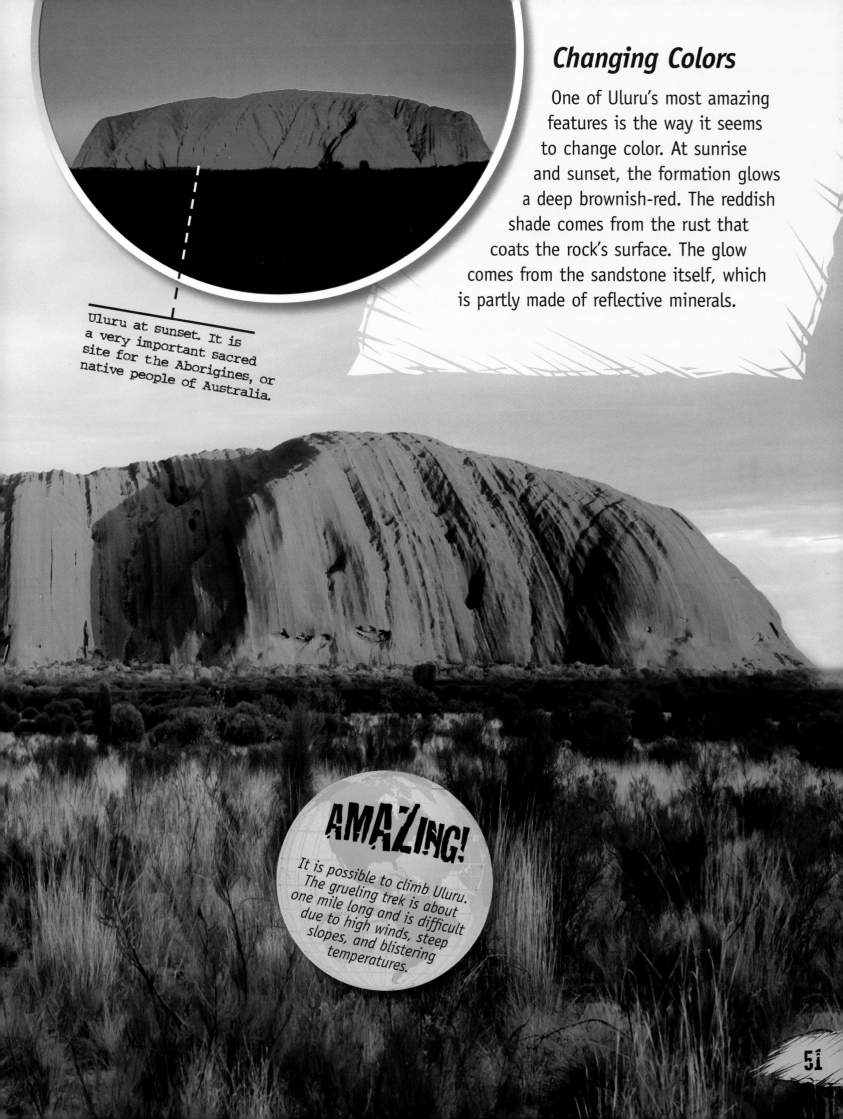

Changing Colors

One of Uluru's most amazing features is the way it seems to change color. At sunrise and sunset, the formation glows a deep brownish-red. The reddish shade comes from the rust that coats the rock's surface. The glow comes from the sandstone itself, which is partly made of reflective minerals.

Uluru at sunset. It is a very important sacred site for the Aborigines, or native people of Australia.

AMAZING!

It is possible to climb Uluru. The grueling trek is about one mile long and is difficult due to high winds, steep slopes, and blistering temperatures.

DEVILS TOWER

The rolling plains of northeastern Wyoming are broken by Devils Tower, a dramatic stone formation that rises almost 1,300 feet above the surrounding terrain. The formation's flattened top is unique and instantly recognized all over the world. The stark sides consist of rocky columns separated by deep cracks. These cracks make Devils Tower a popular destination for climbers, who worm up these broken "tracks" to the rock's summit.

AMAZING!
Devils Tower was the first U.S. National Monument. It was established in 1906 by President Theodore Roosevelt.

Devils Tower is very popular with crack climbers.

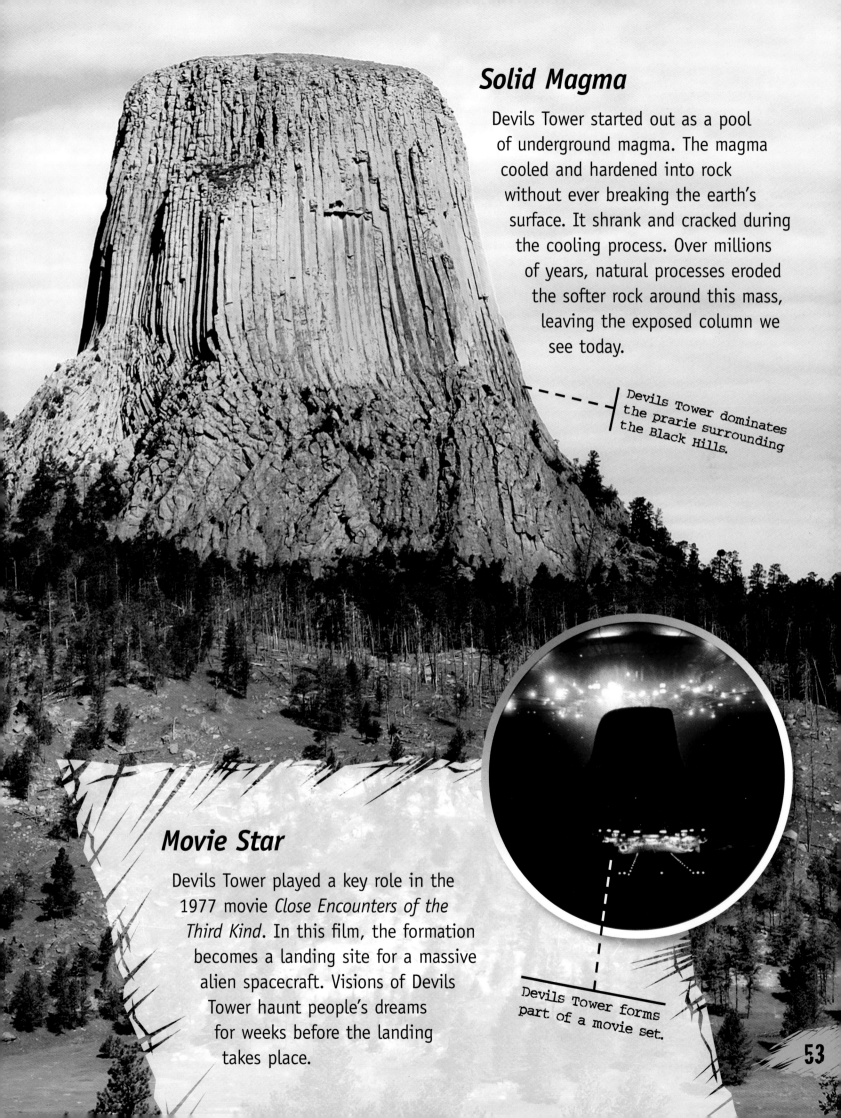

Solid Magma

Devils Tower started out as a pool of underground magma. The magma cooled and hardened into rock without ever breaking the earth's surface. It shrank and cracked during the cooling process. Over millions of years, natural processes eroded the softer rock around this mass, leaving the exposed column we see today.

Devils Tower dominates the prarie surrounding the Black Hills.

Movie Star

Devils Tower played a key role in the 1977 movie *Close Encounters of the Third Kind*. In this film, the formation becomes a landing site for a massive alien spacecraft. Visions of Devils Tower haunt people's dreams for weeks before the landing takes place.

Devils Tower forms part of a movie set.

SHILIN

En masse, the rock formations look like a forest of stone trees.

The naked rocks of Shilin, which means "Stone Forest," erupt from the ground in China's Yunnan Province. The 270-million-year-old formations cover an area of 186 square miles and are made mostly of a type of sedimentary rock called limestone. Along with the stone forests, the area also includes caves, waterfalls, ponds, a lake, and even an underground river.

Shilin is an amazing example of karst topography.

Ashima Stone

Ashima Stone is Shilin's best known and most popular attraction. This rock pillar looks like a girl carrying a basket on her head. According to legend, it formed when a girl named Ashima fled into the forest and turned to stone after being forbidden to marry the man she loved. Local people stage a Torch Festival each year to honor Ashima's memory.

The limestone formation named Ashima Stone at Shilin.

AMAZING!

Shilin's tallest stone pillars rise about 100 feet into the air.

Karst Topography

Shilin is an amazing example of **karst topography**. Karst is a type of landscape that forms when certain rocks, including limestone, are exposed to **acidic** water. The acid eats away the rock over time, leaving odd shapes behind.

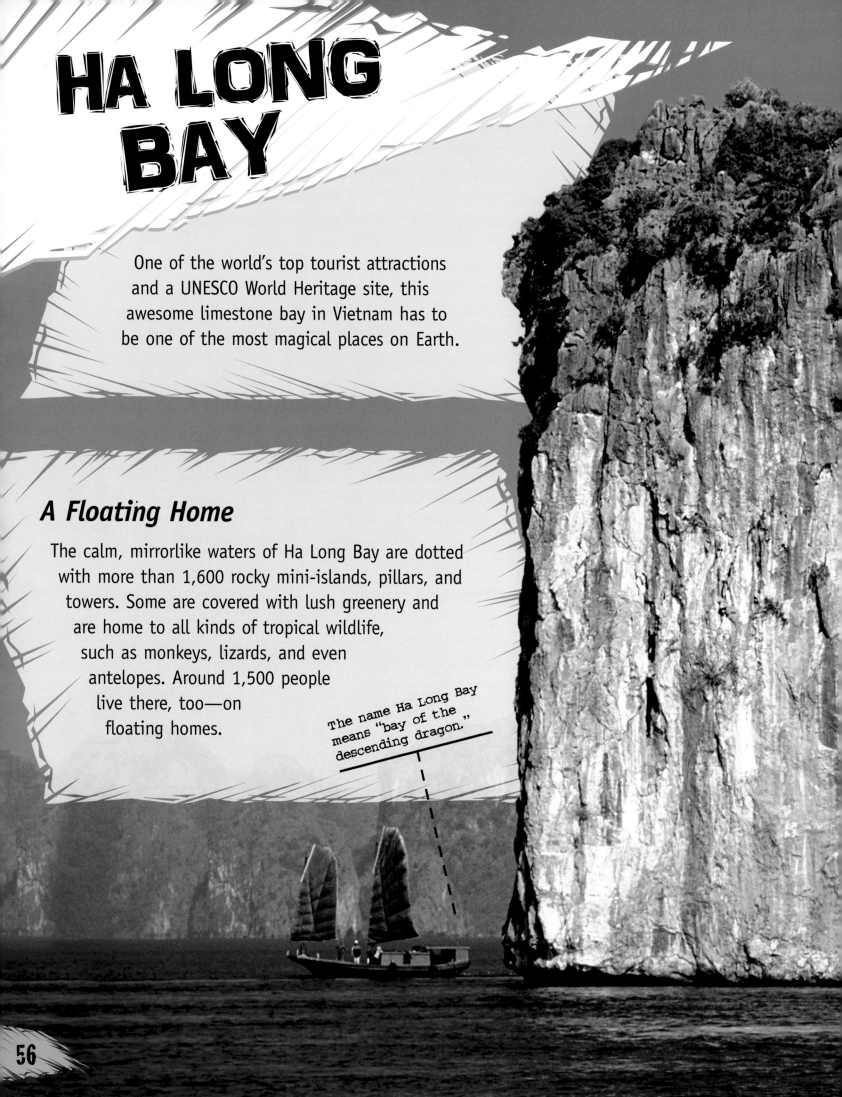

HA LONG BAY

One of the world's top tourist attractions and a UNESCO World Heritage site, this awesome limestone bay in Vietnam has to be one of the most magical places on Earth.

A Floating Home

The calm, mirrorlike waters of Ha Long Bay are dotted with more than 1,600 rocky mini-islands, pillars, and towers. Some are covered with lush greenery and are home to all kinds of tropical wildlife, such as monkeys, lizards, and even antelopes. Around 1,500 people live there, too—on floating homes.

The name Ha Long Bay means "bay of the descending dragon."

Limestone Lumps

Ha Long Bay's islands have taken millions of years to form from limestone that has been shifted, shoved, worn, and dissolved by water and the movements of the earth. Some of the towers are up to 656 feet high, as tall as skyscrapers. Others are hollow lumps with huge, dark caves inside.

Many people take boats to the islands to explore.

AMAZING!

According to legend, Ha Long Bay was formed when dragons spat out jewels, which turned into rocky islands.

GIANT'S CAUSEWAY

The 40,000 **polygonal** pillars of Giant's Causeway blanket a stretch of the Northern Ireland coast, looking like an enormous honeycomb crafted from stone. The pillars formed 50 to 60 million years ago when a molten rock called basalt rose from deep underground. As it cooled, the basalt cracked into mighty columns that remain to this day. The columns vary in height. Some do not break the water's surface, while others are up to 40 feet high.

AMAZING!

Most of the columns at Giant's Causeway are hexagonal, or six-sided, but some columns have four, five, seven, or eight sides.

Built by Giants

Irish folklore claims that Giant's Causeway was the work of a giant named Finn MacCool. MacCool built the causeway as a bridge between Ireland and Scotland after a Scottish giant named Benandonner challenged him to a duel.

Lacada Point

Just next to the Giant's Causeway, there is a rocky finger called Lacada Point. This **promontory** stretches far out to sea and has caused many shipwrecks over the years. The most famous wreck is that of the Spanish galleon *Girona* in 1588. Rediscovered in 1967, the wreck has yielded hundreds of valuable pieces of treasure.

Lacada Point is typical of the rugged yet beautful Antrim coast.

Giant's Causeway is the result of an ancient volcanic explosion.

PRIEKESTOLEN

The world's cliffs were formed millions of years ago by glaciers flowing from the mountains to the sea, and creating deep channels in the rock. Priekestolen, in Norway, is one of these cliffs. This awesome natural formation stands about 2,000 feet above Lysefjord, one of the country's long, deep **fjords**, or sea inlets.

Fear of Heights

For people who suffer from **acrophobia**, or fear of heights, just looking at photos of Priekestolen is probably terrifying. The rock's flat top is bare and windblown—and despite the many thousands of tourists who flock to the site every year, there is no safety rail around the cliff. Accidents are rare, but they have occurred.

It takes between one and three hours to hike to Priekestolen.

Daredevils

Priekestolen is a popular site for daredevils. Each year, thousands of people parachute from the rock or fly off it wearing specially designed wingsuits.

AMAZING!

Priekestolen means "Pulpit Rock." The feature gets this name because it looks a little bit like a huge preacher's pulpit.

Cracked Rock

A deep crack runs across Priekestolen. When the crack gets big enough, the front part of Priekestolen will tumble into the water below. But this will not happen for many, many human lifetimes. For now, this natural wonder is still safe for visitors.

Despite the crack, geologists have confirmed that Priekestolen is safe to visit.

GRAND CANYON

The Grand Canyon, with the Colorado River flowing through it.

The Grand Canyon in Arizona is one of the biggest, most breathtaking rock formations anywhere in the world. This canyon is actually a gorge—a steep-sided river valley. The high cliffs on either side are striped in stunning shades of color, made up of bands of rock of different types and ages. Unlike many other gorges, the Grand Canyon is very wide, varying from 600 feet to 19 miles across. The canyon is nearly 280 miles long and, at its deepest point, is almost 1.25 miles deep.

How Did It Form?

Gorges form when a river cuts through rock as it flows. Over millions of years, the Colorado River has carved its way down through the landscape to create the massive Canyon. The river still flows along the bottom today.

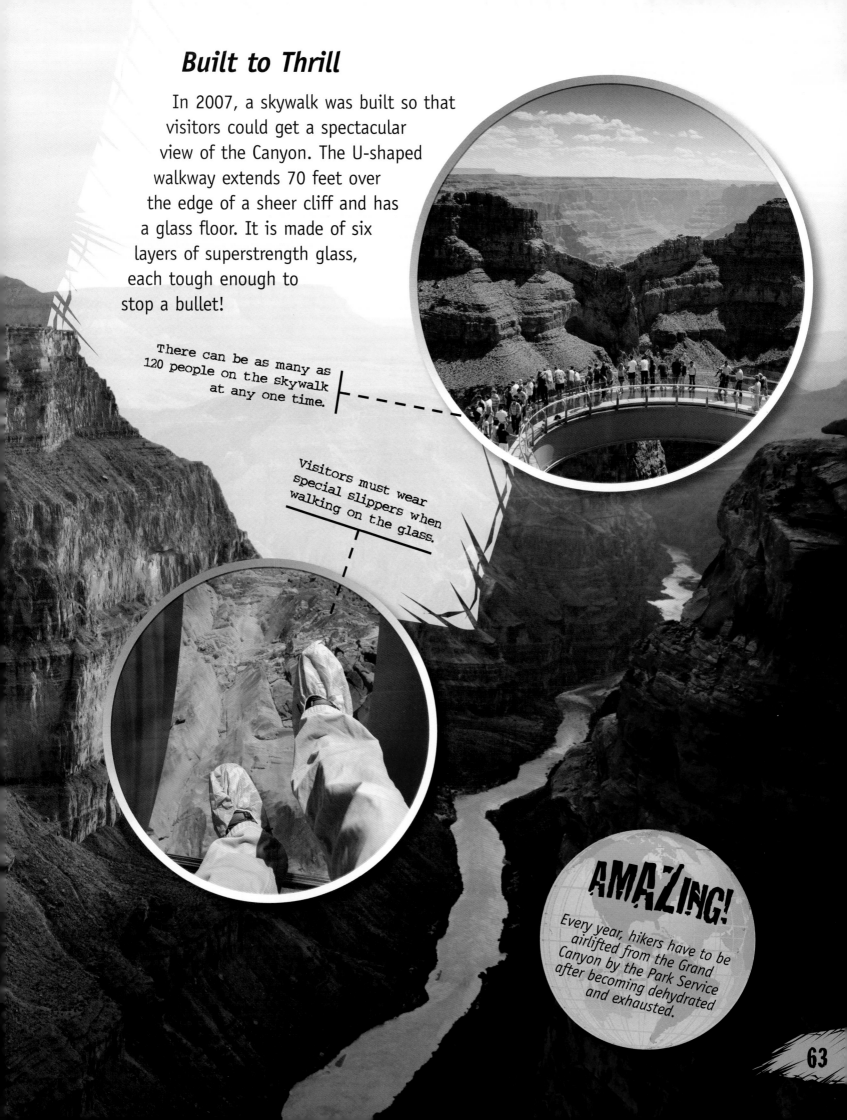

Built to Thrill

In 2007, a skywalk was built so that visitors could get a spectacular view of the Canyon. The U-shaped walkway extends 70 feet over the edge of a sheer cliff and has a glass floor. It is made of six layers of superstrength glass, each tough enough to stop a bullet!

There can be as many as 120 people on the skywalk at any one time.

Visitors must wear special slippers when walking on the glass.

AMAZING!

Every year, hikers have to be airlifted from the Grand Canyon by the Park Service after becoming dehydrated and exhausted.

MAMMOTH CAVE

Mammoth Cave in the state of Kentucky has the longest cave tunnel system in the world. So far, more than 365 miles of linked passageways have been explored on five different levels. And there are undoubtedly more areas to discover—perhaps *many* more. Some scientists believe that the Mammoth Cave system may be more than 1,000 miles long in all.

Cavers in the massive Mammoth Cave.

An underwater lightbulb is replaced in a cave's clear pool.

Underground Life

Parts of Mammoth Cave have underground rivers and pools. In them live several species of troglobites—animals that have adapted to live without sunlight. These shrimp- or crab-like creatures are colorless and eyeless.

Fabulous Features

Mammoth Cave is full of amazing rock formations. Popular areas include Frozen Niagara, which looks like a rocky waterfall; the Drapery Room, where limestone curtains hang from the tall roofs; and the Star Chamber, where crystal chips in the ceiling reflect light to create the illusion of a starry night sky.

You can take a lighted tour to Frozen Niagara.

AMAZING!

Tours of Mammoth Cave have been offered since 1816, making this site one of America's first tourist attractions.

Mining the Cave

In the early 1800s, Mammoth Cave was mined for saltpeter, an ingredient in gunpowder. Most of the gunpowder used in the War of 1812 was made from Mammoth Cave saltpeter. After the war, the mining operation shut down due to lack of demand.

CAVE OF CRYSTALS

Hidden 1,000 feet beneath the desert of northern Mexico lies an astonishing natural treasure: a cave full of massive crystal columns known simply as the Cave of Crystals. Discovered in 2000 during a mining operation, the chamber is crisscrossed with selenite crystals that measure up to 39 feet long and 13 feet wide. The biggest crystals weigh an estimated 55 tons!

Huge crystals dwarf explorers in the Cave of Crystals.

A researcher discovers a pristine crystal.

Extreme Conditions

Visitors to the Cave of Crystals must cope with extreme conditions. The air temperature here can rise to 136°F, and the **humidity** hovers between 90 and 99 percent. Scientists wear ice-packed suits and breathe chilled air so they won't get heatstroke. They also wear goggles to keep their eyeballs from scorching.

Temporary Marvel

Before it was drained in 2000, the Cave of Crystals was filled with water. Today the company that runs the Naica Mine, where the cave is located, keeps the cavity drained. When the draining stops, though—as it will someday—the cave will be flooded once again, and this marvel will be lost.

AMAZING!

It took at least 500,000 years for the mighty columns in the Cave of Crystals to form.

GLOWWORM GROTTO

The ceiling of New Zealand's Glowworm Grotto inside the Waitomo Caves is blanketed with tiny, brilliant blue stars that mimic the most spectacular night sky. These living lights are created inside the bodies of gnat **larvae** that cling to the Grotto's roof. This light, called **bioluminescence**, is caused by a chemical reaction that the gnats can start and stop at will. Unique to New Zealand, these creatures are one of the world's most amazing sights.

Tourists enjoy a dazzling tour of the Glowworm Grotto.

AMAZING!

Glowworm larvae "fish" for prey by dangling sticky threads of mucus from their abdomens.

Glowing Larvae

The gnats of Glowworm Grotto belong to the species *Arachnocampa luminosa*, commonly known as the New Zealand glowworm. The wormlike larvae live for six to twelve months before changing into their adult form. As young larvae, they glow to attract food. As they approach adulthood, the insects may also use their glow to attract mates.

The luminecsent lava of the New Zealand glowworm.

Preservation

Scientists carefully monitor the air quality, temperature, and humidity inside the Waitomo cave system. They have ways to change the air flow patterns inside the caves, if necessary, to keep conditions healthy for the glowworms.

The entrance to the Waitomo Caves.

BRACKEN BAT CAVE

As twilight falls in central Texas, bats start to emerge from the entrance of Bracken Cave. At first, just a few animals appear. As the seconds tick by, though, the trickle turns into a stream—and then into a mighty flood. The sky darkens as more than 20 million bats pour out of the cave and take flight, ready for a night of hunting.

The entrance to Bracken Cave.

Masses of bats emerge from the cave, ready to begin a night of hunting.

Migration

The bats of Bracken Cave form the world's largest bat **colony**. Known as Mexican free-tailed bats, these furry fliers **migrate** 1,000 miles each year from their winter home in Mexico. They live in Bracken Cave from March through October. They head back to Mexico when the weather cools.

Bat Pups

Bracken Cave is the colony's breeding spot. After arriving in the spring, each female bat gives birth to one pup. The pups cling to the ceiling with their tiny claws. They huddle together to share their body heat while their mothers fly outside at night to hunt.

A Mexican free-tailed bat and his pup roosting in the Bracken Bat Cave.

GALÁPAGOS ISLANDS

Almost 600 miles off the coast of Ecuador lie the Galápagos, a group of islands that boasts the world's greatest assortment of wildlife. The incredible diversity of the Galápagos Islands comes partly from the fact that three major ocean currents meet here, drawing thousands of marine species together. It is also partly due to the islands' isolation, which has allowed one-of-a-kind land species to develop and thrive.

Unique Species

The giant tortoise is one of the Galápagos Islands' best known residents. This massive reptile can weigh up to 475 pounds and can live more than 150 years. The Galápagos also host the world's only species of tropical penguins and sea-loving iguanas. Other creatures of note include unique types of sea lions and fur seals.

The giant tortoise is an endangered species.

Darwin's Finches

In 1835, **naturalist** Charles Darwin spent five weeks exploring the Galápagos Islands. Darwin noticed that birds called finches differed slightly from one island to the next. This observation helped to shape Darwin's famous theory of **evolution**, which states that species change slightly over generations as they adapt to their surroundings.

The Darwin Finch, also known as the Galápagos Finch.

AMAZING!

Many Galápagos animals have no natural predators on the islands. For this reason, they are very tame. They let people approach them without fear.

DRAGON ISLANDS

Dragons don't really exist, but there are some animals that come pretty close to these fairy-tale monsters—and they are all found on the "Dragon Islands" in Indonesia. The Komodo dragon, which grows up to ten feet long, is the world's biggest lizard and is found only on the islands of Komodo, Rinca, Flores, Gili Motang, and Padar. Tourists come here from all over the world to marvel at the region's 150-pound monsters.

AMAZING!

Komodo dragons have been known to dig up human graves and eat the decaying bodies inside.

Komodo dragons live on only a few islands in Indonesia, where tourists often go to view them.

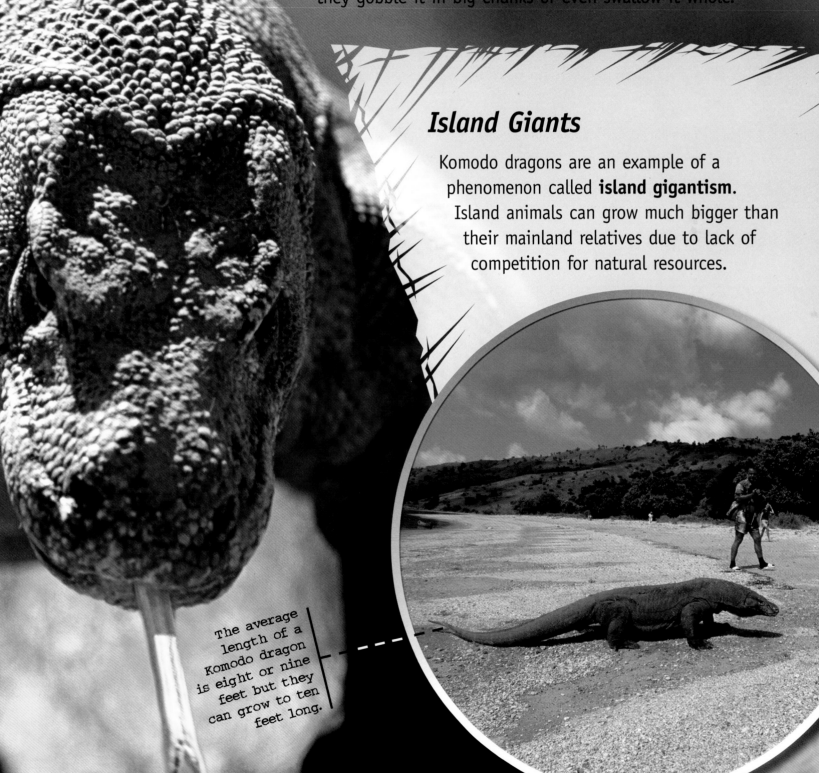

Greedy Guzzler

Komodo dragons have a powerful bite that injects their prey with lethal poison. The poison can cause death in about two days. The dragon then comes back to feast on the dead animal. Komodo dragons can also use their tongues to detect scents in the air, and they can sniff out a rotting carcass from more than five miles away. When they find the source of the smell, such as a dead goat or pig, they gobble it in big chunks or even swallow it whole.

Island Giants

Komodo dragons are an example of a phenomenon called **island gigantism**. Island animals can grow much bigger than their mainland relatives due to lack of competition for natural resources.

The average length of a Komodo dragon is eight or nine feet but they can grow to ten feet long.

GIANT FOREST

Covering an area of barely three square miles, the Giant Forest might not seem all that enormous—until you look up! As the world's largest grove of giant sequoias, which regularly grow to heights of 250 feet with trunk diameters of 25 feet or more, this patch of forest is truly remarkable. It is the centerpiece of California's Sequoia National Park and home to five of our planet's biggest known trees.

The drive through tunnel log at Sequoia National Park.

There's a Museum in the Giant Forest where you can find out more about sequoias.

76

General Sherman

A sequoia named General Sherman is not only the largest tree in Giant Forest; by weight, it is the largest living **organism** on Earth! General Sherman is 275 feet tall and has a trunk circumference of 102 feet. It is estimated to weigh 2.7 million pounds.

The General Sherman: the largest tree in the world!

Old Age

Giant sequoias are one of three species of redwoods. They are not the tallest redwoods, but they are the thickest and most massive. They can grow to at least 3,500 years of age and are among the longest-lived organisms on Earth.

Scientists look at cross sections of sequoias to age the trees.

LOCH NESS

Loch Ness is Scotland's second largest body of water by surface area and its largest by volume. It is famous as the home of Nessie, a monster that allegedly lives in the Loch's murky depths. Nessie was first "seen" in 1933. Since then, there have been more than 3,000 reported sightings of this beast that looks, according to those who have spotted it, like some kind of prehistoric water reptile with a long, thin neck and a tiny head.

The ruined Urquhart Castle overlooks Loch Ness.

AMAZING!

A 2014 satellite photo of Loch Ness appeared to show a 100-foot creature wading underwater. Could it be Nessie?

Is there really a monster living in the depths of Loch Ness?

The Surgeon's Photograph

The most famous image of Nessie is called the Surgeon's Photograph. It was snapped in 1934 by doctor Robert Kenneth Wilson. Although the photo was later uncovered as a hoax, believers say that this fact does not disprove the monster's existence.

The famous image of Nessie taken in 1934.

Mythical Beast

The Loch Ness Monster is an example of a **cryptid**. Cryptids are animals whose existence has been suggested by the public, but which have not been verified by the scientific community. Besides Nessie, other well-known cryptids include Bigfoot, mermaids, the goat-eating chupacabra, and the Yeti.

79

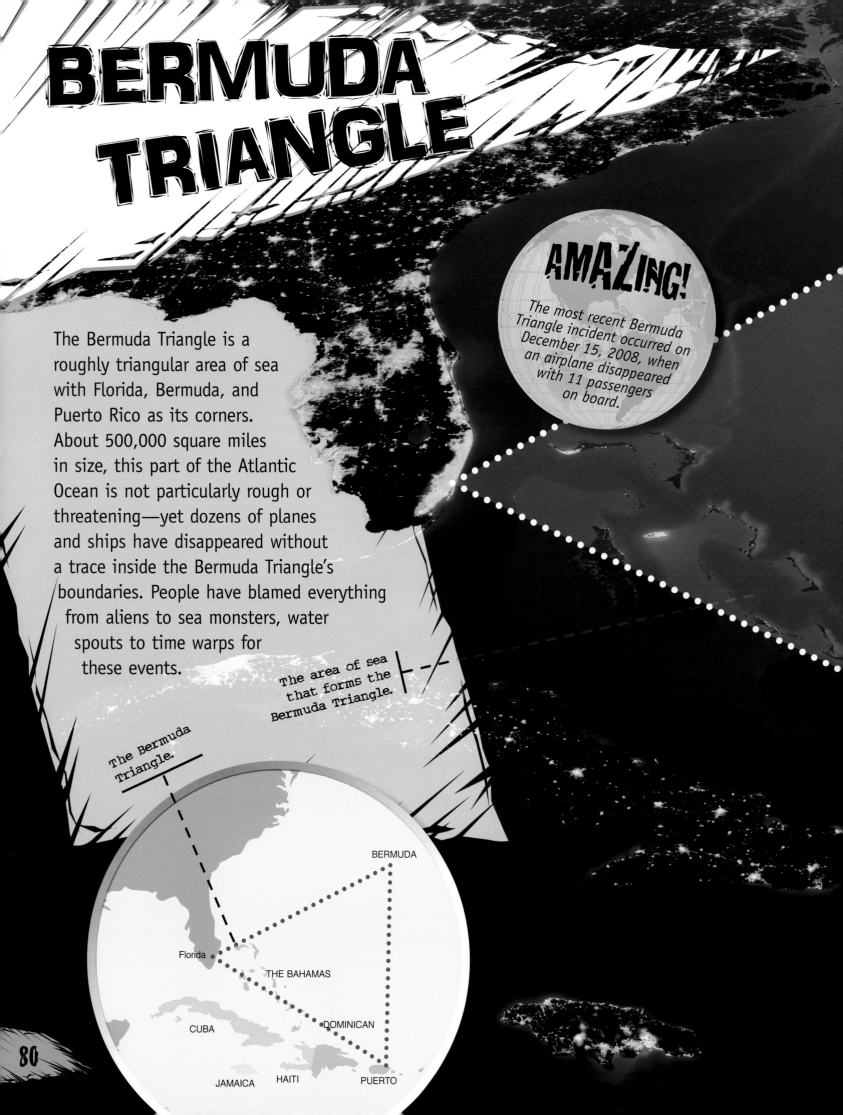

BERMUDA TRIANGLE

The Bermuda Triangle is a roughly triangular area of sea with Florida, Bermuda, and Puerto Rico as its corners. About 500,000 square miles in size, this part of the Atlantic Ocean is not particularly rough or threatening—yet dozens of planes and ships have disappeared without a trace inside the Bermuda Triangle's boundaries. People have blamed everything from aliens to sea monsters, water spouts to time warps for these events.

AMAZING!

The most recent Bermuda Triangle incident occurred on December 15, 2008, when an airplane disappeared with 11 passengers on board.

The area of sea that forms the Bermuda Triangle.

The Bermuda Triangle.

BERMUDA

Florida

THE BAHAMAS

CUBA

DOMINICAN

JAMAICA HAITI PUERTO

Flight 19

On December 5, 1945, a training flight of five bombers took off from Fort Lauderdale, Florida, and headed out over the Bermuda Triangle's waters. The bombers were supposed to make a short circle over the sea, then return to base—but the airplanes were never heard from again. Search and rescue vehicles were sent out to find the missing Flight 19 but one of these ships, too, vanished without a trace.

The legendary lost squadron from Flight 19.

Busy Area

Some of the world's busiest airline and shipping lanes cross the Bermuda Triangle. Experts say that the number of incidents in the Triangle is normal considering the amount of traffic in the area.

WRECK OF THE TITANIC

Off the northeastern coast of Canada, the North Atlantic Ocean is stormy, freezing cold, and dotted with icebergs. These conditions proved to be the undoing of the RMS *Titanic*, a ship that sank in these waters after striking an iceberg on April 15, 1912. The wreck of the *Titanic* lies more than two miles below the water's surface on the sea floor, where it has been visited and studied by many scientific expeditions.

RMS Titanic in 1912.

Lost and Found

Although many people looked for the *Titanic* in the years after she sank, the ship remained lost for decades. The wreck was finally located in 1985 by an expedition run by scientist Robert Ballard. A camera-loaded sled, towed back and forth underwater, captured the first images of the sunken ship.

The bow section of Titanic, discovered at the bottom of the ocean in 1985.

Debris Field

A **debris** field surrounds the wreck of the *Titanic*. Covering an area of about two square miles, the debris field contains ship parts, suitcases, clothing, wine bottles, bathtubs, and other items ejected from the sinking ship.

Debris such as this lifering was still being found months after the disaster.

STONEHENGE

What is the purpose of Stonehenge, an ancient and massive stone monument located on a chalky plain in England? No one knows for sure—but people have plenty of guesses. Some people say the stone circles were used in religious rituals or for healing purposes. Others believe Stonehenge is a primitive observatory where long-ago people studied the skies. Some people even think Stonehenge has magical properties.

An aerial view of Stonehenge.

Shape and Size

The structure called Stonehenge today was probably built between 4,000 and 5,000 years ago. It consists of two rings of rocks. The upright rocks, called **sarsens**, are about 30 feet tall. They are set in pairs. Smaller horizontal rocks called **lintels** lie across the top of each pair of sarsens.

Bluestones

Smaller rocks called bluestones are scattered throughout Stonehenge. These rocks weigh up to 8,000 pounds each. They were transported to Stonehenge thousands of years ago from Wales, a distance of about 140 miles. Ancient people must have thought the bluestones had special properties to go to this effort.

Stonehenge is a UNESCO world heritage site.

The bluestones were transported from Wales to the stonehenge site in Salisbury, Hampshire.

AMAZING!

There are dozens of burial mounds and shrines near Stonehenge. Thousands of bodies were cremated or buried here.

EASTER ISLAND

In 1722, Dutch explorer Jacob Roggeveen landed at a tiny volcanic island in the Pacific Ocean. As it was Easter Day, he named the island Easter Island. The island was dotted with enormous statues of humans up to 72 feet tall. Known as **moai**, these awesome sculptures may have been made as a form of ancestor worship. However, no one knows for sure how or why the Easter Island moai were constructed.

The moai line the shore like rocky sentinels.

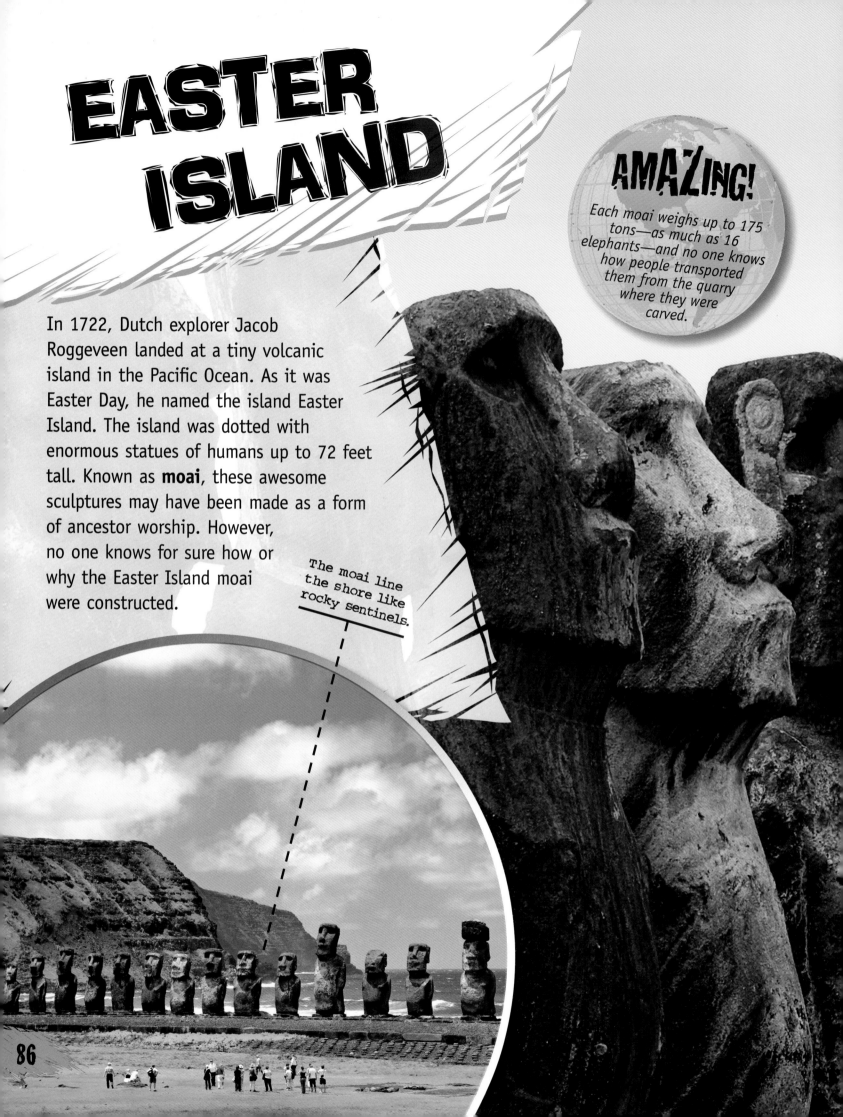

Humongous Heads

The moai are often called the Easter Island Heads. They are whole body statues, but their heads are very large, and often the lower part of each statue is buried. Over the years most of the statues were damaged, but many have now been restored.

Moai dot the banks of an extinct volcano on Easter Island.

The famous Easter Island Heads.

Isolated Island

According to a 2012 census, 5,761 people live on Easter Island. This remote spot lies more than 2,000 miles off the west coast of South America, and it is about 1,300 miles away from the nearest inhabited island. With these stats, Easter Island qualifies as one of the world's most isolated places to live.

NAZCA LINES

The Nazca Lines are a collection of hundreds of pictures and patterns cut into the ground. The images are so vast that they only make sense when seen from the sky. They are found on a high, flat stretch of the Peruvian Desert in South America, close to the town of Nazca. These images are the most famous examples of formations called **geoglyphs**—patterns or pictures created on the ground by manipulating elements of the landscape.

Some of the pictures are hundreds of feet across. This one is of a monkey.

Long-Lasting Lines

Experts think the Nazca Lines were made between 1,500 and 2,000 years ago. Early Nazca people created them by lifting a layer of red stones on the surface of the desert to reveal the whiter rock underneath. The patterns have survived because there is little rain or wind in the desert to disturb them.

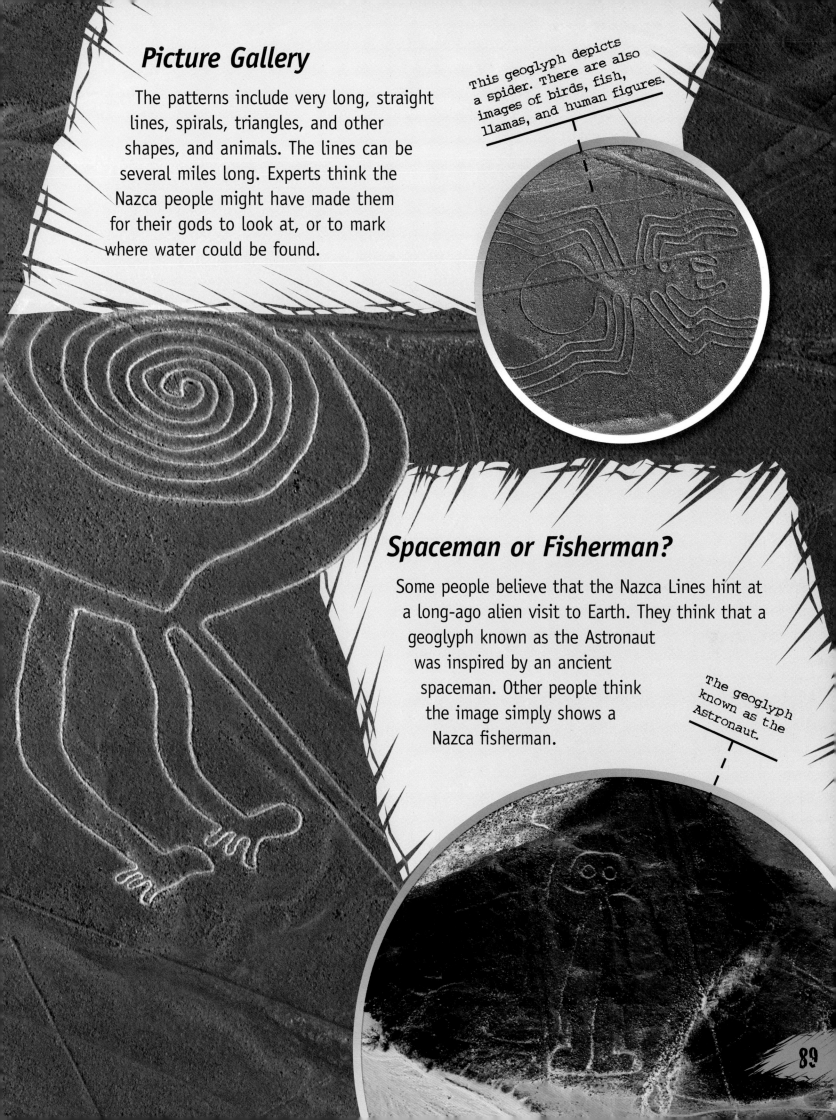

Picture Gallery

The patterns include very long, straight lines, spirals, triangles, and other shapes, and animals. The lines can be several miles long. Experts think the Nazca people might have made them for their gods to look at, or to mark where water could be found.

This geoglyph depicts a spider. There are also images of birds, fish, llamas, and human figures.

Spaceman or Fisherman?

Some people believe that the Nazca Lines hint at a long-ago alien visit to Earth. They think that a geoglyph known as the Astronaut was inspired by an ancient spaceman. Other people think the image simply shows a Nazca fisherman.

The geoglyph known as the Astronaut.

LASCAUX CAVES

In 1940, a young Frenchman named Marcel Ravidat was walking in the woods when he came upon an opening in the ground, caused by a toppled tree. Ravidat entered the hole and found a cave that had walls decorated with vivid images of charging bulls, leaping stags, and other beasts. Ravidat had discovered the Lascaux Caves, a site that boasts the world's best-preserved collection of prehistoric artwork.

The beasts shown in the art probably lived in the caves.

Old Art

The 2,000 images in the Lascaux Caves are more than 17,000 years old. They depict animals, humans, and abstract signs. The most famous part of the cave is the Great Hall of the Bulls, which includes a 17-foot-long painting of a bull—the largest animal discovered so far in cave art.

The Great Hall of the Bulls.

AMAZING!

The original Lascaux Caves are now closed to visitors. However, an exact copy of the Caves, known as Lascaux II, is open to the public.

Preservation

The opening of the Lascaux Caves has taken a toll on the paintings inside. Changes in air movement and light conditions inside the cave, along with the breath of 1,200 visitors per day, have caused lichens and crystals to grow on the walls. Most recently, a black fungus has begun to threaten the prehistoric paintings. Scientists are fighting to cure this blight and protect the treasures of Lascaux.

PALERMO CATACOMBS

Underground tunnels dug by humans can be hiding places, scenes for religious ceremonies, or places for burying the dead. In Palermo, Italy, **catacombs** were used to store dead bodies. This happened when the Capuchin Monastery outgrew its original cemetery in the 16th century. The monks built crypts below the monastery to store more bodies.

One of the mummified bodies in the catacombs at Palermo.

The catacombs contain around 1,252 mummies.

Mummified Monk

The Palermo catacombs contain **mummies**—preserved bodies with skin, flesh, and hair. In 1599, a monk was buried in the vault under the monastery. Weirdly, the monk's body did not rot away. When people heard about this, many wanted to be buried there.

What Makes a Mummy?

It is not magic that makes the Palermo catacombs good at preserving bodies. It is the presence in the soil of a rock called tufa that soaks up moisture. In a very dry atmosphere, things don't rot easily, so this helps keep the mummies fresh.

Uncannily lifelike mummies in the Palermo catacombs.

Pay or Move!

The Palermo catacombs were maintained through donations. Relatives of the deceased were expected to send cash on a regular basis. If they did not do this, the body of their loved one was moved from its regular place and stuck on a back shelf until payment resumed.

POMPEII

Ancient Pompeii was a thriving town near modern-day Naples, Italy. The city sat in the shadow of an active volcano called Mt. Vesuvius. In 62 CE, much of Pompeii was destroyed by an earthquake. The earthquake was caused by **magma** inside Mt. Vesuvius. The magma was under great pressure. It shook the ground as it tried to find a way to the surface.

Volcanic Eruption

Over many years, the pressure built and built—and on August 24, 79 CE, the magma finally broke free. Rock, ash, and lava erupted into the air. Clouds of ash rose high into the sky, blocking out the sunlight. This material began to settle onto Pompeii and its neighboring towns.

AMAZING!

Mt. Vesuvius is still active. Its most recent eruption occurred in 1944. Scientists consider it one of the world's most dangerous volcanoes.

The ruins of Pompeii as they look today, with Mount Vesuvius looming in the background.

This painting shows the eruption of Mount Vesuvius in 1760—one of many eruptions in the eighteenth century.

Cast in Stone

The citizens of Pompeii hid indoors and curled into balls, trying to protect themselves. But there was no escape. The ash smothered thousands of people and hardened around the victims' bodies, forming human-shaped cavities. In 1860, when the city was first excavated, casts were made of these cavities with plaster of Paris. The casts show people moments from death. They remain today as evidence of one of history's greatest natural disasters.

Some of the casts of human bodies at Pompeii.

Tourists explore the ruins in Pompeii—it's a fascinating place to visit.

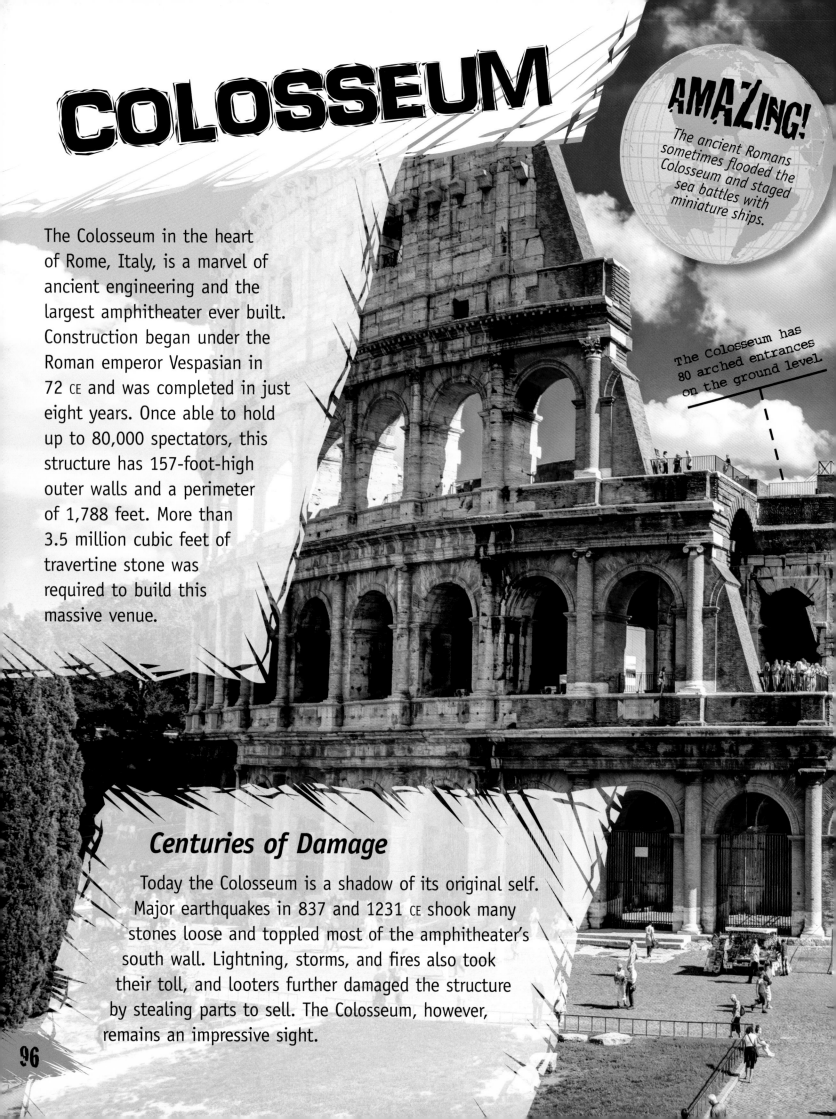

COLOSSEUM

The Colosseum in the heart of Rome, Italy, is a marvel of ancient engineering and the largest amphitheater ever built. Construction began under the Roman emperor Vespasian in 72 CE and was completed in just eight years. Once able to hold up to 80,000 spectators, this structure has 157-foot-high outer walls and a perimeter of 1,788 feet. More than 3.5 million cubic feet of travertine stone was required to build this massive venue.

The Colosseum has 80 arched entrances on the ground level

Centuries of Damage

Today the Colosseum is a shadow of its original self. Major earthquakes in 837 and 1231 CE shook many stones loose and toppled most of the amphitheater's south wall. Lightning, storms, and fires also took their toll, and looters further damaged the structure by stealing parts to sell. The Colosseum, however, remains an impressive sight.

Brutal Entertainment

The Colosseum was used mostly to stage battles of different types. Some battles were between people and others were between dangerous animals, for example a tiger fighting an elephant or a bear fighting a lion. Fights also took place between unarmed humans and animals. Huge crowds gathered to watch these events, which were fought to the death.

Professional fighters called gladiators fought at the Colosseum.

The inside of the Colosseum as it looks today.

MACHU PICCHU

Machu Picchu has been a UNESCO World Heritage Site since 1981.

Perched nearly 8,000 feet above sea level on a mountaintop in Peru is Machu Picchu, a city built by the Incas in the 1400s. The city was probably built as an estate for the Inca emperor Pachacuti, although no one knows this for sure. We do know that this city in the clouds was fully self-contained and produced all of its own food and water—enough to support a population of more than 1,000 people.

Skilled Architects

The Incas were skilled architects, and the 150-plus buildings of Machu Picchu showcase this talent. The buildings' walls are made of stones chiseled to fit together precisely, like the pieces of a jigsaw puzzle, without the need for mortar. The joints are so tight that even today, more than 500 years after these walls were built, a thin knife blade cannot fit between the stones.

A close-up of the Artisans' Wall shows how tightly packed the stones are.

98

Terraces

The Incas leveled the mountaintop where Machu Picchu rests by carving **terraces** into the hillsides. There are about 700 terraces in the city, and they provide flat areas for farmland, lawns, streets, and much more.

Agricultural terraces formed part of the Machu Pichu complex.

AMAZING!

Machu Picchu was abandoned less than 100 years after it was built. Historians suspect that the residents died of smallpox.

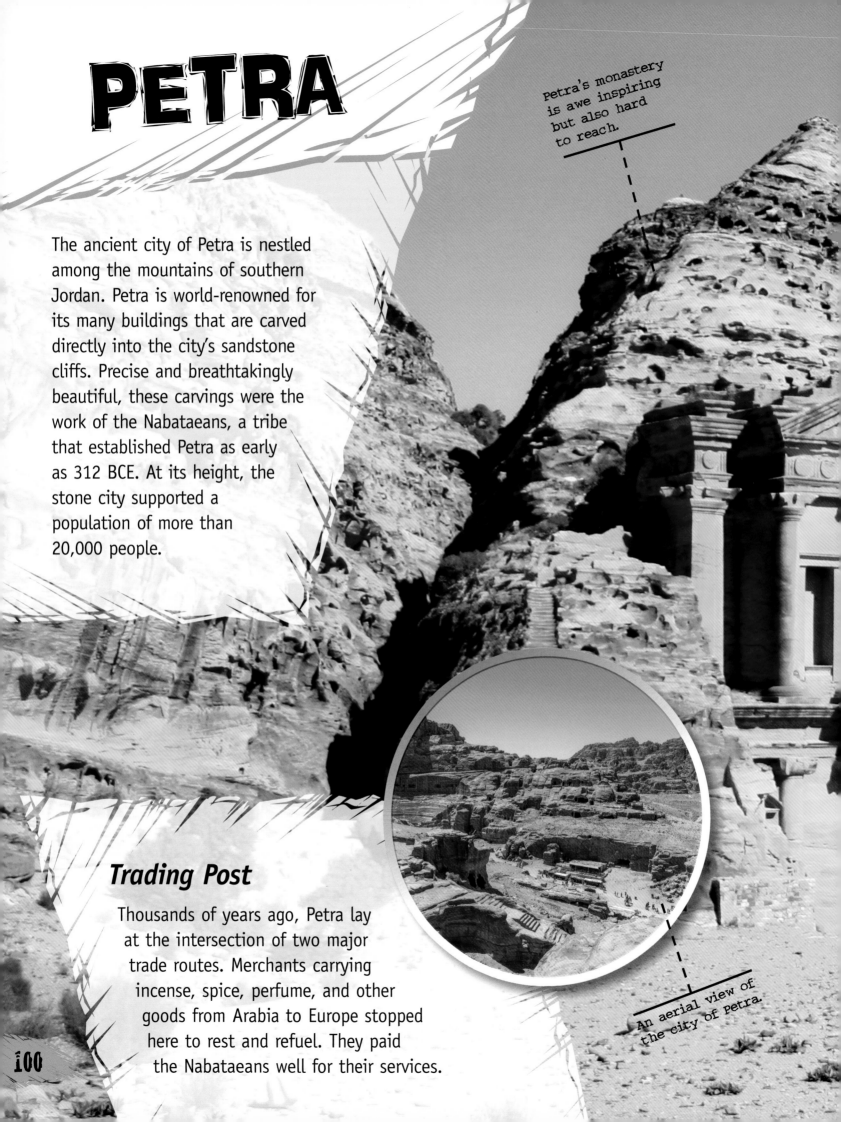

PETRA

Petra's monastery is awe inspiring but also hard to reach.

The ancient city of Petra is nestled among the mountains of southern Jordan. Petra is world-renowned for its many buildings that are carved directly into the city's sandstone cliffs. Precise and breathtakingly beautiful, these carvings were the work of the Nabataeans, a tribe that established Petra as early as 312 BCE. At its height, the stone city supported a population of more than 20,000 people.

Trading Post

Thousands of years ago, Petra lay at the intersection of two major trade routes. Merchants carrying incense, spice, perfume, and other goods from Arabia to Europe stopped here to rest and refuel. They paid the Nabataeans well for their services.

An aerial view of the city of Petra.

Bigger is Better

Thanks to the merchant trade, Petra and its residents were very wealthy—and they loved to display their riches. Commissioning elaborate, expensive carvings on their homes was one way to do this. Making enormous structures was another. From the Nabataean viewpoint, bigger was always better.

Dwellings were carved into the stone cliffs of ancient Petra.

AMAZING!

Even though it sat in the desert, ancient Petra had a sophisticated water storage and distribution system that could supply up to 100,000 people.

ANGKOR WAT

Angkor Wat, the world's largest religious monument, covers an astonishing area of 401 acres. It was built in the early 1100s under the direction of Cambodia's King Suryavarman II. It was originally designed to be three things: a Hindu temple, the state capital, and the king's eventual **mausoleum**. By the late 1200s, Angkor Wat had become a Buddhist temple instead, and this use continues to the present day.

An aerial view of Angkor Wat.

Sacred Mountain

Angkor Wat was designed to imitate Mount Meru, a mythical sacred mountain in both Hindu and Buddhist traditions. The temple's five central towers represent Mount Meru's five peaks. The walls around the temple represent the mountain ranges around Mount Meru, and Angkor Wat's moat represents the distant sea.

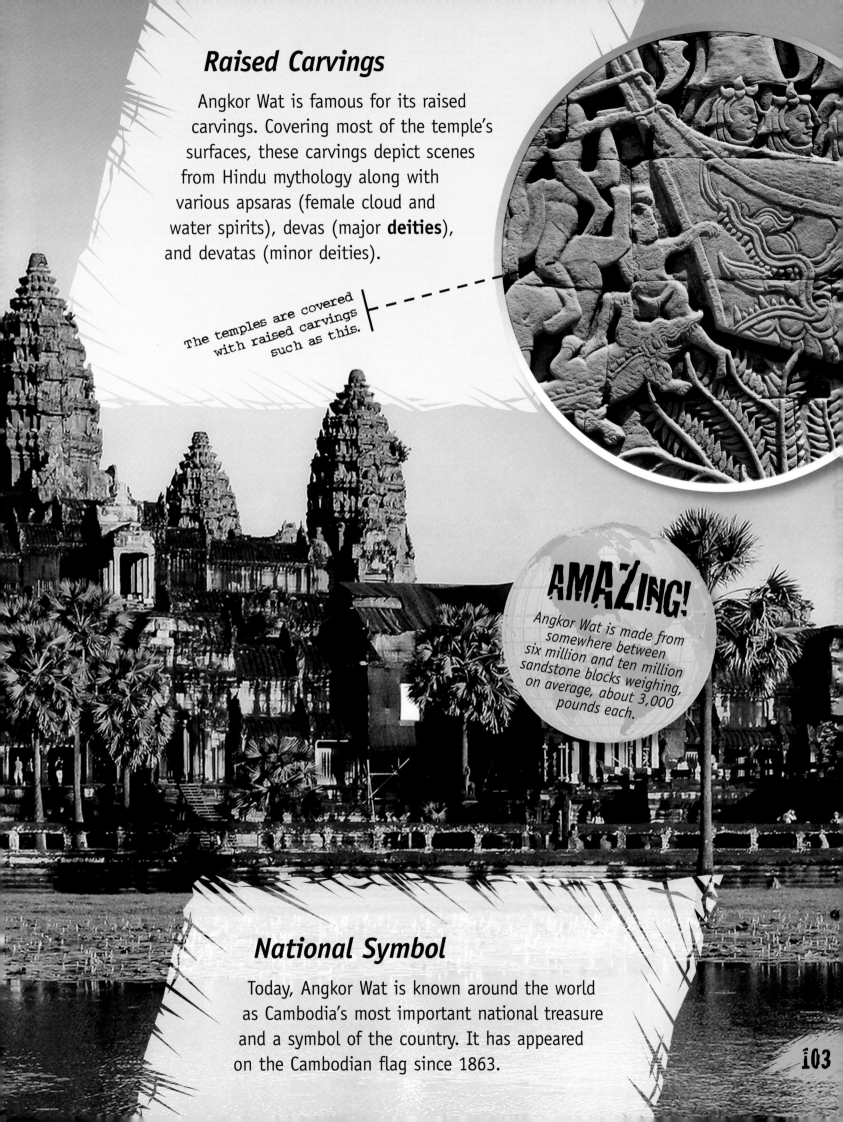

Raised Carvings

Angkor Wat is famous for its raised carvings. Covering most of the temple's surfaces, these carvings depict scenes from Hindu mythology along with various apsaras (female cloud and water spirits), devas (major **deities**), and devatas (minor deities).

The temples are covered with raised carvings such as this.

AMAZING!

Angkor Wat is made from somewhere between six million and ten million sandstone blocks weighing, on average, about 3,000 pounds each.

National Symbol

Today, Angkor Wat is known around the world as Cambodia's most important national treasure and a symbol of the country. It has appeared on the Cambodian flag since 1863.

103

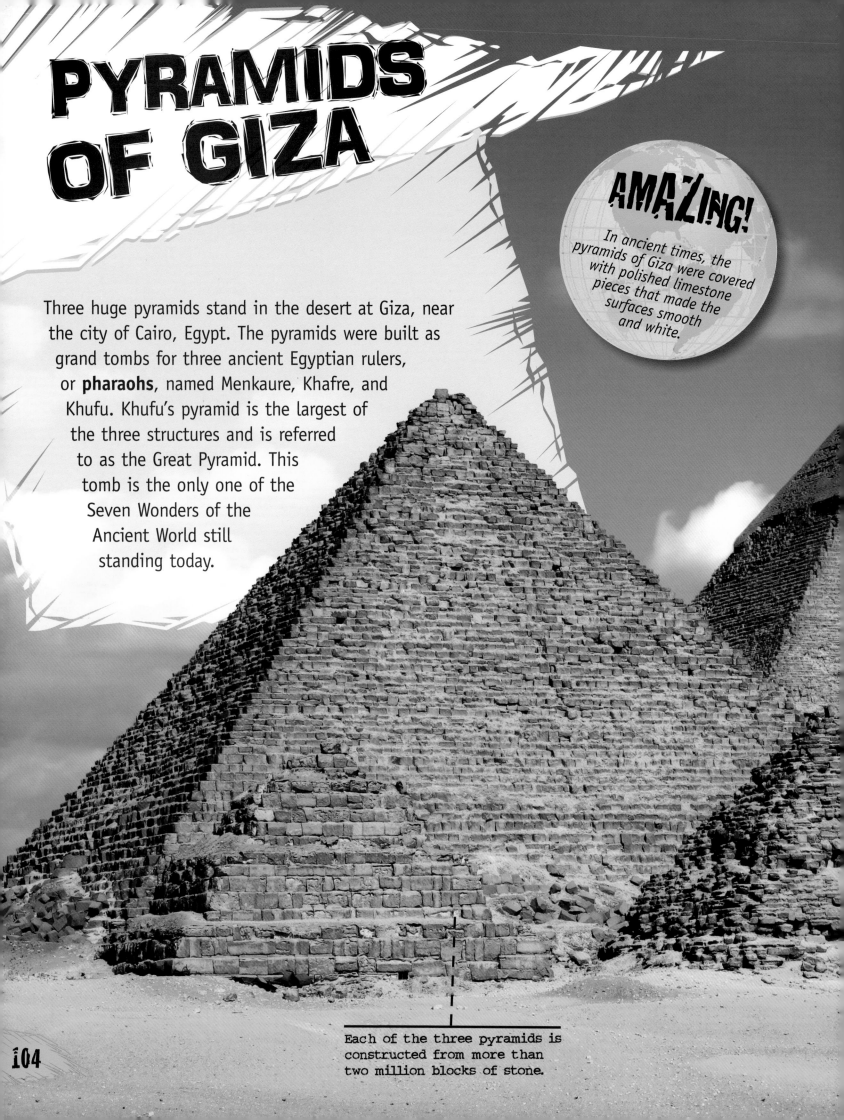

PYRAMIDS OF GIZA

Three huge pyramids stand in the desert at Giza, near the city of Cairo, Egypt. The pyramids were built as grand tombs for three ancient Egyptian rulers, or **pharaohs**, named Menkaure, Khafre, and Khufu. Khufu's pyramid is the largest of the three structures and is referred to as the Great Pyramid. This tomb is the only one of the Seven Wonders of the Ancient World still standing today.

Each of the three pyramids is constructed from more than two million blocks of stone.

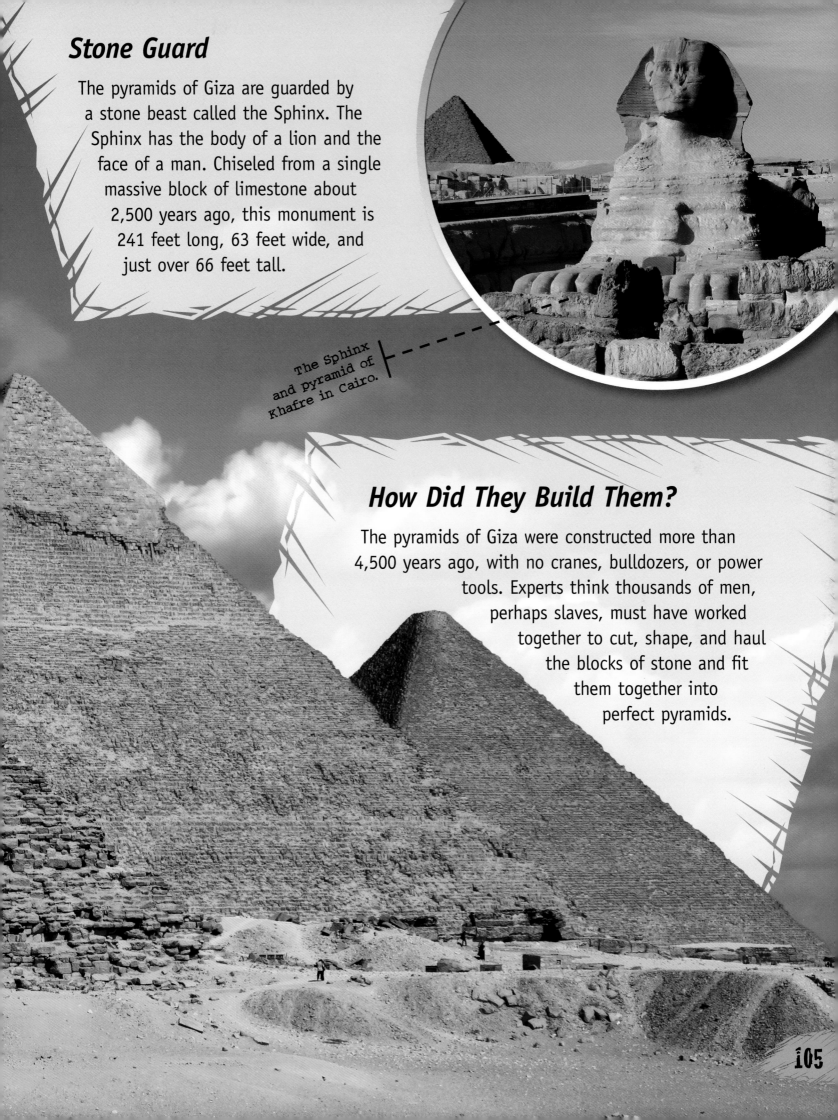

Stone Guard

The pyramids of Giza are guarded by a stone beast called the Sphinx. The Sphinx has the body of a lion and the face of a man. Chiseled from a single massive block of limestone about 2,500 years ago, this monument is 241 feet long, 63 feet wide, and just over 66 feet tall.

The Sphinx and Pyramid of Khafre in Cairo.

How Did They Build Them?

The pyramids of Giza were constructed more than 4,500 years ago, with no cranes, bulldozers, or power tools. Experts think thousands of men, perhaps slaves, must have worked together to cut, shape, and haul the blocks of stone and fit them together into perfect pyramids.

GREAT WALL OF CHINA

Stretching across northern China for roughly 5,500 miles is the Great Wall of China. Built from stone, brick, and soil, the Wall once protected the Chinese Empire from invaders. It stands today as one of history's most impressive architectural feats.

The Jinsanling section of the Great Wall in the mountanous Luanping area.

Keep Out!

People began building walls across northern China 2,600 years ago. In 214 BCE, Emperor Shi Huangdi ordered the connection of existing walls and also had new walls built to keep out hostile neighbors. Most of these original walls have now worn away. The Great Wall as it is today was mostly built in the 1400s and 1500s to keep out Mongolian invaders.

Emperor Shi Huangdi.

AMAZING!

If you walked along China's Great Wall for eight hours a day, it would take you seven months to reach its end.

Mass Grave

The men who built the Great Wall during Shi Huangdi's time were mostly soldiers and prisoners. The work was brutal, and many people—up to 400,000, according to some estimates—died during the process. Many of the workers' bodies were tossed into the Wall's foundation, which became a mass grave as construction continued.

TAJ MAHAL

More than 20,000 workers and 1,000 elephants helped build the Taj Mahal, a tomb for the wife of the Indian emperor. When Mumtaz Mahal died in 1631, her husband, Shah Jahan, commissioned a beautiful tomb in her memory. The Taj Majal is made from white marble inlaid with 28 precious and semiprecious stones.

The Taj Mahal is in Agra, north India. It attracts thousands of visitors every day.

AMAZING!

In the war years of 1942, 1965, and 1971, the Indian government built wooden scaffolding around the Taj Mahal to hide it from enemy bombers.

Mourning for Mumtaz

No expense was spared in the building of the tomb. Materials came from all over Asia, and the top designers, sculptors, mosaic artists, and calligraphers formed the creative team. Mumtaz Mahal was buried there once the 22-year building process was complete, and Shah Jahan joined her when he died in 1666.

Reports say that when Mumtaz died, Shah Jahan's grief was so great that his black hair turned white.

Under Threat

Today the Taj Mahal is being damaged by acidic rain, which eats away the tomb's marble walls and turns the rock yellow. To slow this damage, the Indian government has created a 4,000-square-mile zone around the monument where strict **emissions** standards are in place. Controlling emissions reduces the acid level of the rain.

NEUSCHWANSTEIN CASTLE

Neuschwanstein Castle was the pet project of Ludwig II, king of Bavaria, from 1864 to 1886. This fairytale structure perches atop a craggy hill near the village of Fussen in southern Germany. Blending Romantic design with medieval touches, Neuschwanstein was meant to be Ludwig II's personal estate and private retreat. However, the project was not yet complete when the king died suddenly at the age of 40.

The luxurious walls of the castle are paintings from medieval legends.

Inspired by Opera

Neuschwanstein Castle was partly inspired by Ludwig II's love of opera, and by the work of composer Richard Wagner in particular. Murals showing scenes from Wagner's works decorate the walls inside the castle. The castle's name, which means "New Swan Stone" in English, refers to a swan knight from the opera *Lohengrin*.

The Castle just before sunrise.

Built For Looks

Ludwig II had a very specific vision of the way his castle should look. To achieve this look, the king hired a theatrical stage designer instead of an architect to plan Neuschwanstein Castle. The designer came up with many of the elaborate but useless touches that give the building so much charm.

The Sleeping Beauty Castle at Disneyland.

STATUE OF LIBERTY

The Statue of Liberty stands on Liberty Island in New York Harbor. The 151-foot figure depicts Libertas, the Roman goddess of liberty, bearing a stone tablet and holding a torch aloft. Designed by French sculptor Frédéric Auguste Bartholdi and installed in 1886, the statue was a gift to America from France. "Lady Liberty," as she is often called, was built to celebrate the end of the American Civil War and the **abolition** of slavery in the United States.

The statue was built in Paris, broken into parts, packed into huge crates, and shipped to New York.

Why Green?

The surface layer of the Statue of Liberty is made of thin copper sheets. The copper was originally a dull brown color. As the copper was exposed to the air, it developed a greenish coating called verdigris. This coating does not harm the statue; in fact, it protects it from the elements.

AMAZING!

Lady Liberty has a 35-foot waistline. She wears size 879 shoes, and her index finger is 8 feet long!

The original torch was replaced in 1984 by a copper torch covered in 24 karat gold leaf.

Climbing the Statue

More than 3 million tourists visit the Statue of Liberty each year. Of these people, only 240 per day—10 per group, three groups per hour—are allowed to enter Lady Liberty and climb into her crown. Visitors buy their tickets for this hard-to-get tour up to a year in advance.

Visitors must climb 354 stairs to reach Lady Liberty's crown!

MOUNT RUSHMORE

The faces of four great U.S. presidents from history—George Washington, Thomas Jefferson, Theodore Roosevelt, and Abraham Lincoln—are carved into the grey granite cliffs of South Dakota. Called Mount Rushmore, this mighty monument is known today as one of America's national symbols and attracts millions of visitors each year.

The monument is carved into the peaks of the Black Hills in South Dakota.

Long Life

Granite is a hard rock that erodes very slowly. The carvings of Mount Rushmore are expected to wear away at a rate of about one inch per 10,000 years. With regular cleaning to remove lichen and other growths, the carvings could last for hundreds of thousands of years.

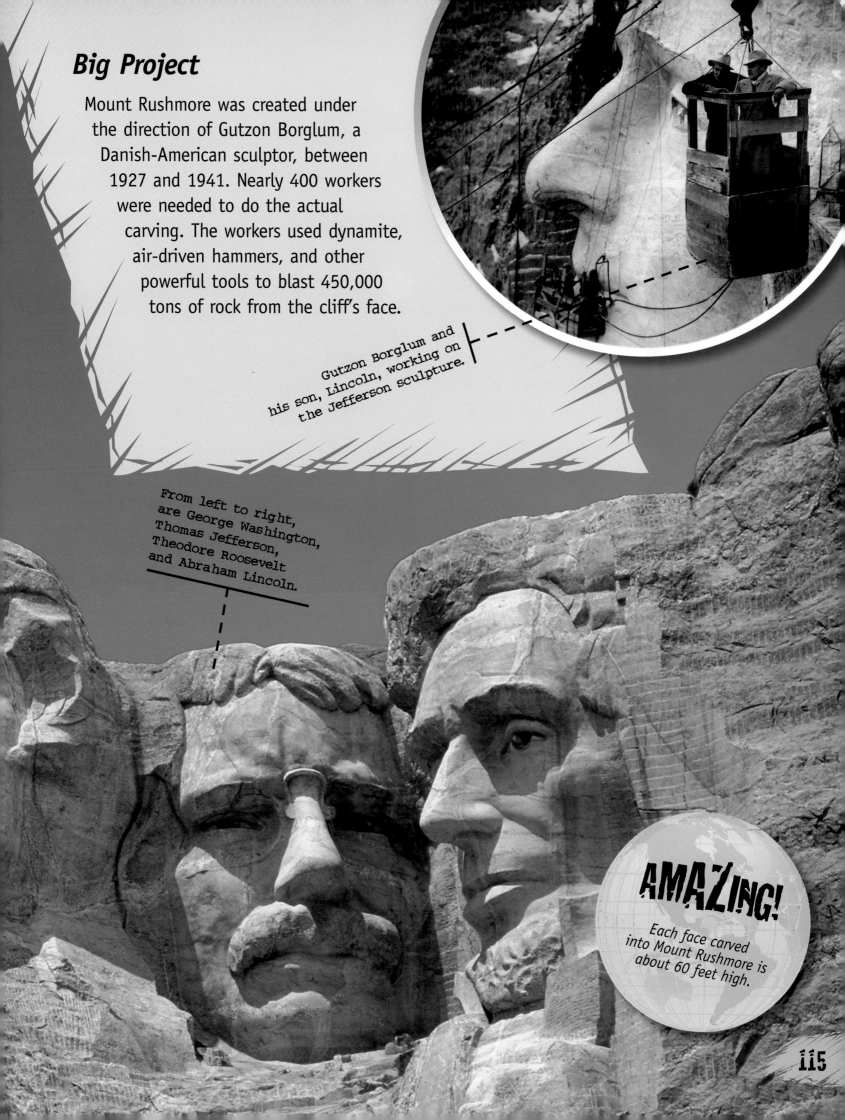

Big Project

Mount Rushmore was created under the direction of Gutzon Borglum, a Danish-American sculptor, between 1927 and 1941. Nearly 400 workers were needed to do the actual carving. The workers used dynamite, air-driven hammers, and other powerful tools to blast 450,000 tons of rock from the cliff's face.

Gutzon Borglum and his son, Lincoln, working on the Jefferson sculpture.

From left to right, are George Washington, Thomas Jefferson, Theodore Roosevelt and Abraham Lincoln.

AMAZING!

Each face carved into Mount Rushmore is about 60 feet high.

FLOATING ISLANDS

Hundreds of years ago, the Uros people of Peru were forced out of their homes by the expanding Inca empire. In desperation, the Uros built floating islands of reeds to live on and launched themselves into Lake Titicaca. They built a life there that has endured to the present day. Today there are about 60 floating islands in Lake Titicaca, built and maintained by a population of about 1,200 people.

Totora Reeds

The floating islands are made from **totora** reeds, which are native to Lake Titicaca. The reeds are woven into islands up to 12 feet thick. The residents' houses, boats, and even beds are also made of totora reeds. The islands are anchored to the lake floor and can be moved if necessary.

Totora reeds, shown here on Lake Titicaca, also grow on Easter Island in the Pacific Ocean.

Modern Amenities

Despite their traditional lifestyle, the Uros people enjoy modern amenities. They use solar panels to run televisions and other electrical appliances. They have reed schools for their children. The largest floating island even has a radio station that broadcasts for several hours each day.

Solar power is used to power electrical appliances.

The Uros people build islands and their homes from reeds.

AMAZING!

The floating islands need constant repair. Old reeds rot away and must be replaced with new ones.

VENICE

The city of Venice, Italy, sits on 117 small islands in northeastern Italy. The islands are separated by canals and linked by an extensive bridge system. The canals act as roads for boats, which are the city's only motorized vehicles. The bridges let the city's 270,000 residents—and up to 20 million tourists annually—walk from one solid area to another.

For centuries gondoliers were the main method of transport in Venice. You can still enjoy a gondola ride today.

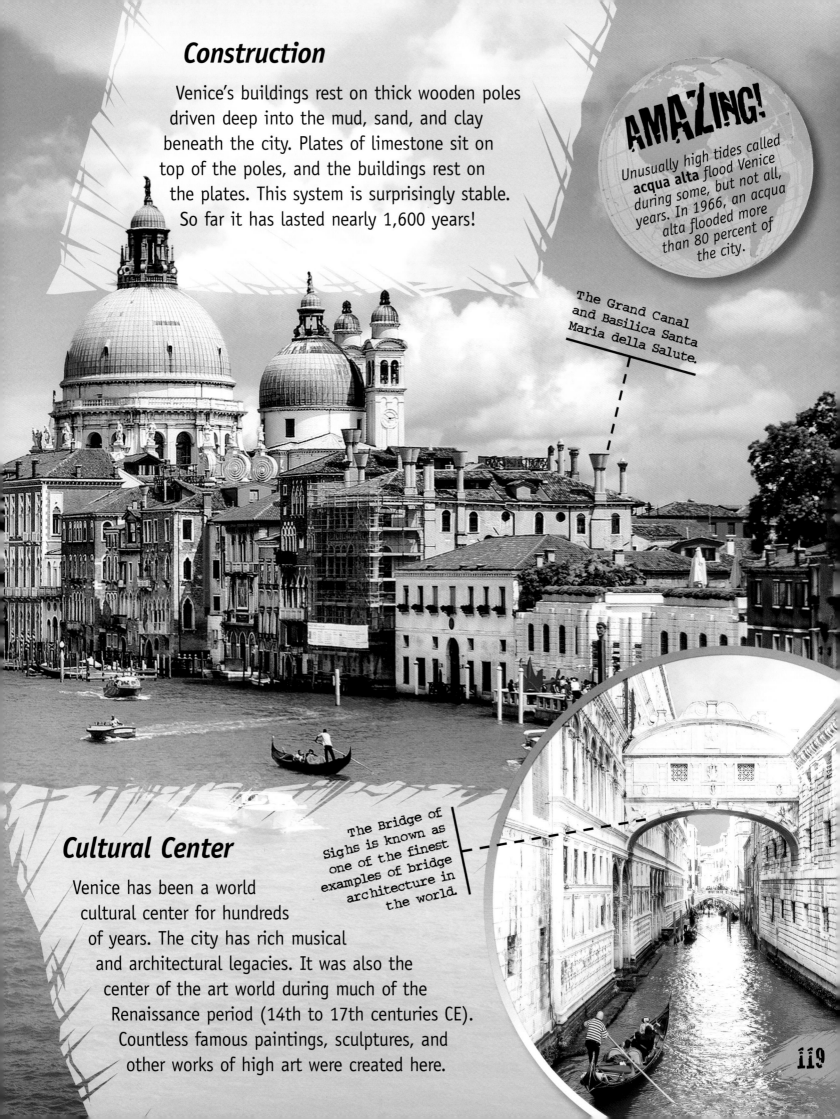

Construction

Venice's buildings rest on thick wooden poles driven deep into the mud, sand, and clay beneath the city. Plates of limestone sit on top of the poles, and the buildings rest on the plates. This system is surprisingly stable. So far it has lasted nearly 1,600 years!

AMAZING!

Unusually high tides called **acqua alta** flood Venice during some, but not all, years. In 1966, an acqua alta flooded more than 80 percent of the city.

The Grand Canal and Basilica Santa Maria della Salute.

Cultural Center

The Bridge of Sighs is known as one of the finest examples of bridge architecture in the world.

Venice has been a world cultural center for hundreds of years. The city has rich musical and architectural legacies. It was also the center of the art world during much of the Renaissance period (14th to 17th centuries CE). Countless famous paintings, sculptures, and other works of high art were created here.

EIFFEL TOWER

The Eiffel Tower in Paris, France, was completed in 1889. It was originally built as the entrance to the World's Fair, which was held in Paris that year to mark the 100-year anniversary of the French Revolution. More than 100 artists submitted plans for the monument. The winning design came from a construction firm belonging to Alexandre-Gustave Eiffel, whose name became permanently attached to the structure his firm built.

AMAZING!

The Eiffel Tower is coated with more than 50 tons of paint every seven years to protect it from rust.

Tall Structure

The final design of the Eiffel Tower called for more than 18,000 iron girders and 2.5 million rivets. Several hundred workers spent two years assembling the Tower's framework. When complete, the Eiffel Tower was the tallest building in the world, with an elevation of 1,063 feet. Today, many taller structures exist in the world—but the Eiffel Tower is still France's second-highest manmade structure.

The Eiffel Tower during construction. Just the base and first platform have been built at this point.

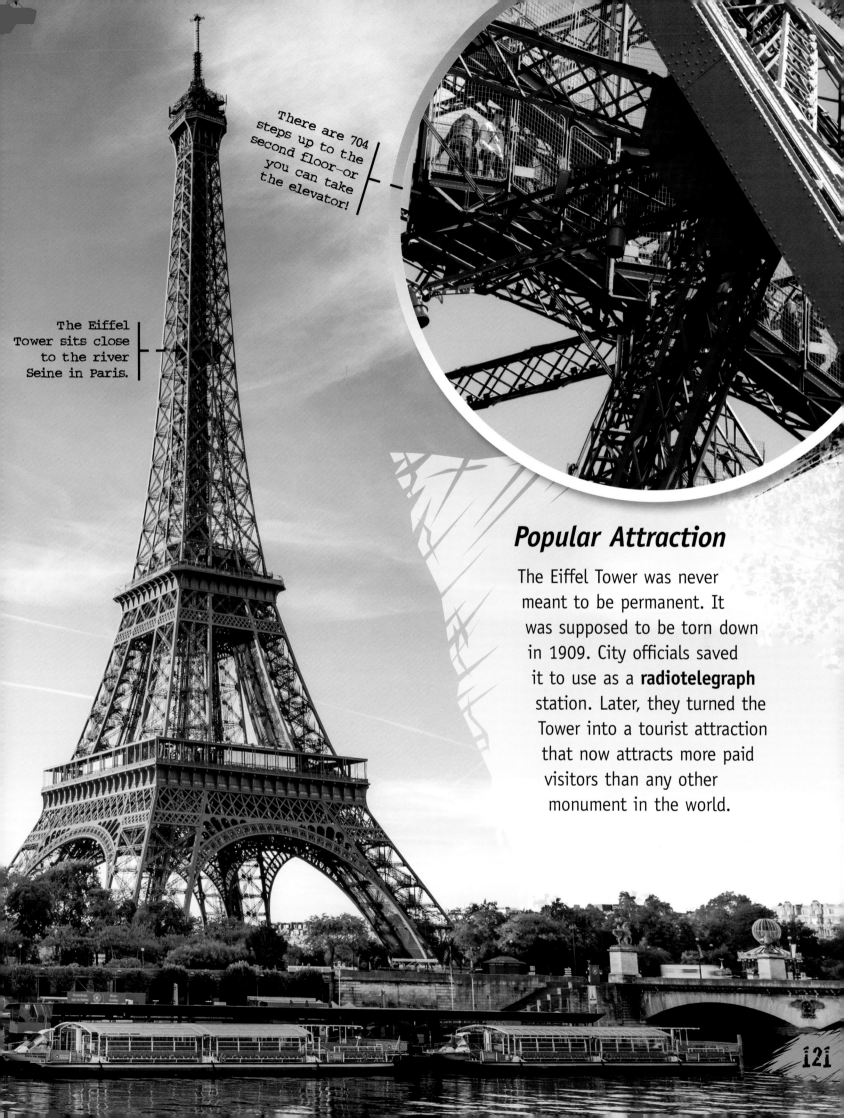

There are 704 steps up to the second floor—or you can take the elevator!

The Eiffel Tower sits close to the river Seine in Paris.

Popular Attraction

The Eiffel Tower was never meant to be permanent. It was supposed to be torn down in 1909. City officials saved it to use as a **radiotelegraph** station. Later, they turned the Tower into a tourist attraction that now attracts more paid visitors than any other monument in the world.

MONT SAINT-MICHEL

During periods of low tide, the mound-shaped city of Mont Saint-Michel sits on a vast marshy plain off the coast of Normandy, France. As the tide rises, however, water pours across the muddy ground in great, rolling waves. It flows around the walled city, cutting it off from the land. The galloping liquid rises up to 46 feet, turning Mont Saint-Michel into an island until the tide turns and the water flows out again.

Mont Saint-Michel city at sunset.

The interior of the abbey that sits at the top of Mont Saint-Michel.

The Abbey

A great abbey sits atop Mont Saint-Michel. Built in the 16th century, the abbey towers over a small but once thriving village. At its peak, Mont Saint-Michel had a population of more than 1,000 people. Today about 50 monks still live on the mound. They are caretakers for the monument and also hold Mass in the abbey each day.

Mont Saint-Michel sits about half a mile from land.

Prison Rock

Mont Saint-Michel was used as a prison for several decades in the 1800s. The fast, deadly tidal changes made this facility almost impossible to escape.

LA JUMENT

The sea that borders Brittany, a region in northwestern France, is stormy and dangerous. It is also an important lane for shipping traffic. Brittany's coastline is dotted with lighthouses that guide these ships away from deadly rocks. La Jument, a lighthouse off the island of Ushant, is the best known of these amazing structures.

La Jument lighthouse as seen from Ushant island.

The lighthouse during stormy weather.

Lighthouse keeper
Theodore Malgorne
peers from the
lighthouse doorway.

Famous Photo

On December 21, 1989, a raging
storm was battering La Jument.
Photographer Jean Guichard flew to the lighthouse
in a helicopter to take pictures of the spectacle.
Lighthouse keeper Theodore Malgorne came outside
when he heard the helicopter—and was almost
washed away by a massive wave. Guichard caught
the perilous moment on film, creating one of the
most famous lighthouse images ever taken.

AMAZING!

In La Jument's first year of
construction, sea conditions
were so bad that only
52 hours of work could
be completed.

Wave-Washed

Lighthouses built in the water, like La Jument,
are called wave-washed lighthouses. Keepers
once lived in all of these lighthouses. Today,
thanks to technological advances, most of
them are unmanned. La Jument has been
fully **automated** since 1991.

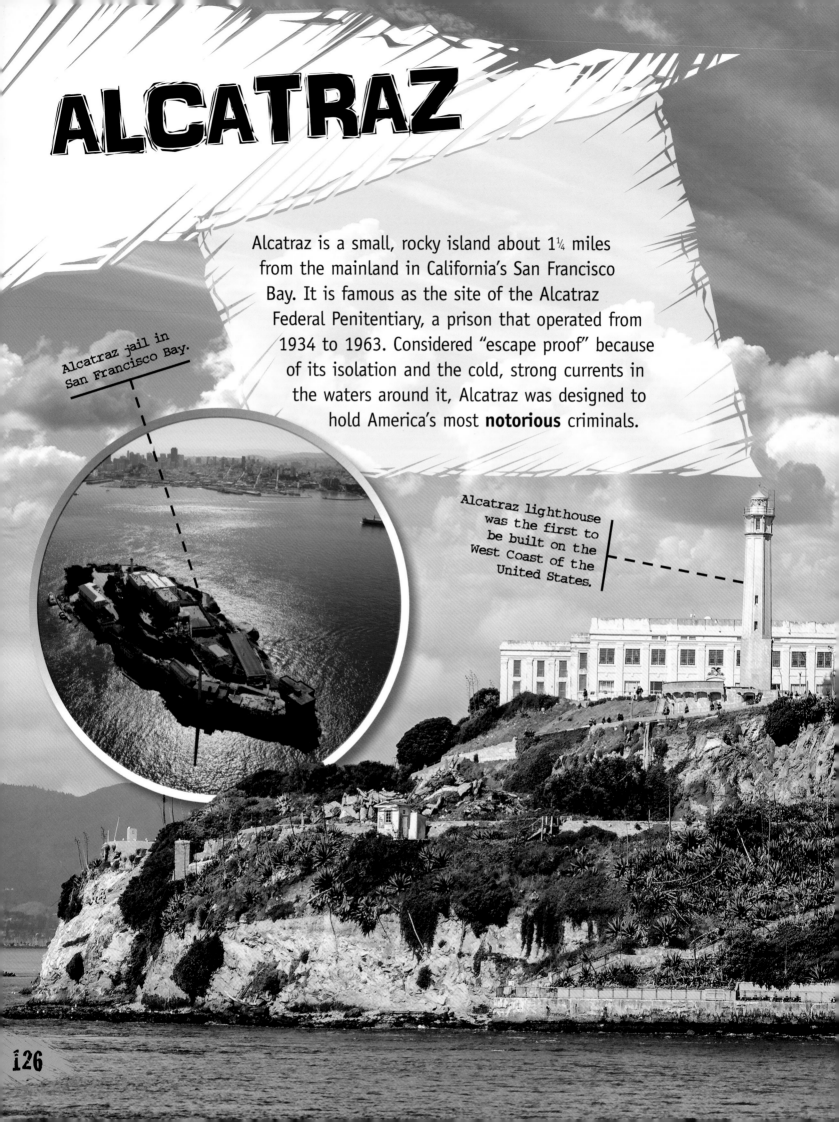

ALCATRAZ

Alcatraz is a small, rocky island about 1¼ miles from the mainland in California's San Francisco Bay. It is famous as the site of the Alcatraz Federal Penitentiary, a prison that operated from 1934 to 1963. Considered "escape proof" because of its isolation and the cold, strong currents in the waters around it, Alcatraz was designed to hold America's most **notorious** criminals.

Alcatraz jail in San Francisco Bay.

Alcatraz lighthouse was the first to be built on the West Coast of the United States.

Escape Attempts

A total of 36 prisoners made 14 escape attempts during Alcatraz's years of operation. Of these men, 23 were caught alive, six were shot and killed during their escape, two drowned, and five are listed as missing and presumed drowned.

The hospital at Alcatraz included several wards with beds.

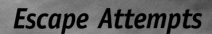

AMAZING!

No executions took place at Alcatraz. However, there were five suicides and eight murders during the prison's 29 years of operation.

Shut Down

Per inmate, the Alcatraz Federal Penitentiary cost three times as much to run as any other U.S. **penal** institution. It was shut down in 1963 for budget reasons. It reopened in 1973 as a museum and tourist attraction. Today, the prison, its surrounding buildings, and Alcatraz Island attract more than 1.3 million visitors each year.

FORBIDDEN CITY

Panoromic view of the Forbidden City.

At the heart of Beijing lies the Forbidden City, a 178-acre palace complex that housed China's emperors from 1420 to 1911 CE. Only the emperors, their families, their servants, and government officials were allowed inside the complex during this nearly 500-year period. All others were forbidden to enter on pain of death, a rule that gave the compound its name.

Structure

The Forbidden City contains about 980 buildings, including everything from small dwellings to soaring audience halls. Rectangular in shape, the City is surrounded by 32-foot-high walls and a 171-foot-wide, water-filled moat. Four bridges (one on each side) provide access to the area. The largest bridge leads to the mighty Meridian Gate, which is the main entrance into the City.

The moat leads past the Meridian Gate and the entrance to the City.

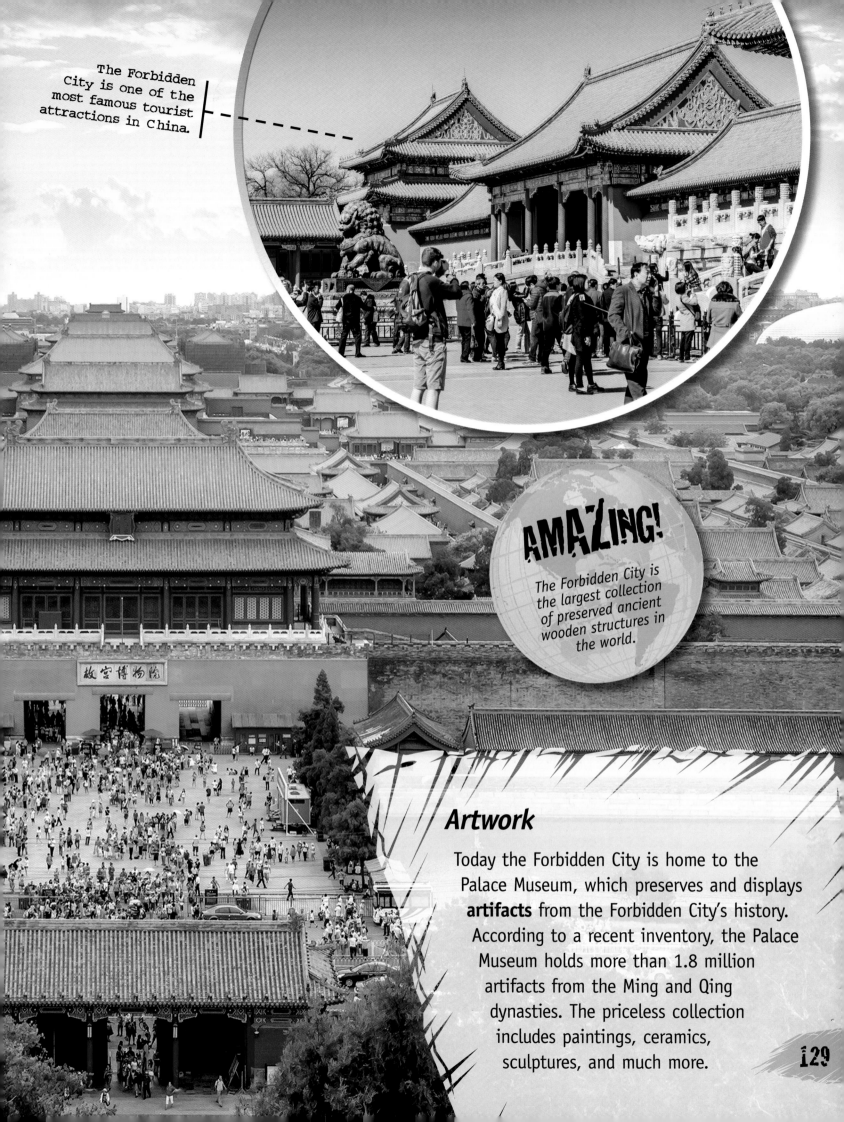

The Forbidden City is one of the most famous tourist attractions in China.

AMAZING!

The Forbidden City is the largest collection of preserved ancient wooden structures in the world.

故宫博物院

Artwork

Today the Forbidden City is home to the Palace Museum, which preserves and displays **artifacts** from the Forbidden City's history. According to a recent inventory, the Palace Museum holds more than 1.8 million artifacts from the Ming and Qing dynasties. The priceless collection includes paintings, ceramics, sculptures, and much more.

AKASHI KAIKYO BRIDGE

Since 1998, when it was opened to traffic, the Akashi Kaikyo Bridge in Japan has held the title of the world's longest suspension bridge. A suspension bridge is a type of bridge that uses towers and cables to support the weight of its roadway, or **deck**. The main deck of the Akashi Kaikyo Bridge is 6,532 feet long. Adding the bridge's two other deck sections of 3,150 feet each, the structure's total length is a staggering 12,832 feet long. That's more than two miles!

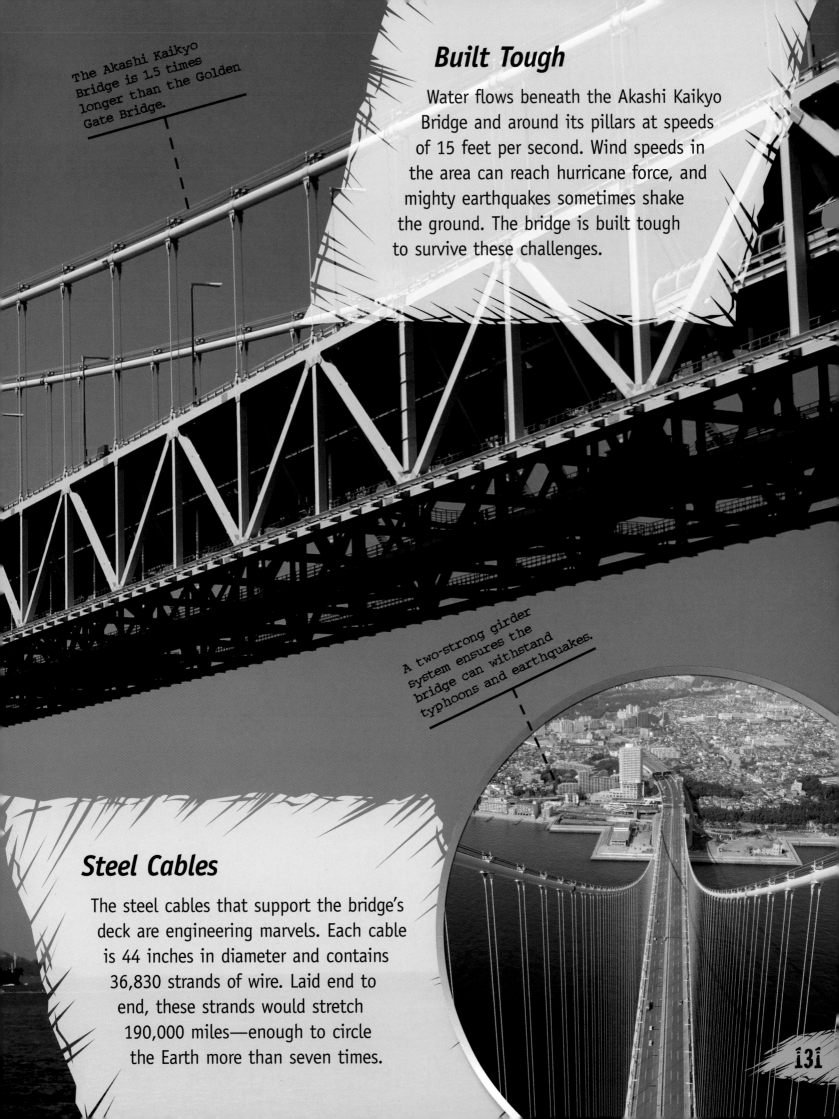

The Akashi Kaikyo Bridge is 1.5 times longer than the Golden Gate Bridge.

Built Tough

Water flows beneath the Akashi Kaikyo Bridge and around its pillars at speeds of 15 feet per second. Wind speeds in the area can reach hurricane force, and mighty earthquakes sometimes shake the ground. The bridge is built tough to survive these challenges.

A two-strong girder system ensures the bridge can withstand typhoons and earthquakes.

Steel Cables

The steel cables that support the bridge's deck are engineering marvels. Each cable is 44 inches in diameter and contains 36,830 strands of wire. Laid end to end, these strands would stretch 190,000 miles—enough to circle the Earth more than seven times.

ICEHOTEL

The ICEHOTEL in Jukkasjärvi, Sweden, is quite literally the world's coolest hotel! This 65-room lodging is built from scratch each winter using ice blocks from the nearby Torne River. The hotel's walls, floors, ceilings, furniture, and even plates and cups are made of ice. Guests sleep on frozen beds covered with thick reindeer pelts in rooms where the temperature never rises above 23°F.

AMAZING!

Each spring, the ICEHOTEL melts completely. A few key parts are kept in freezer storage for the following year.

Reindeer pelts insulate the doors that lead into the ICEHOTEL.

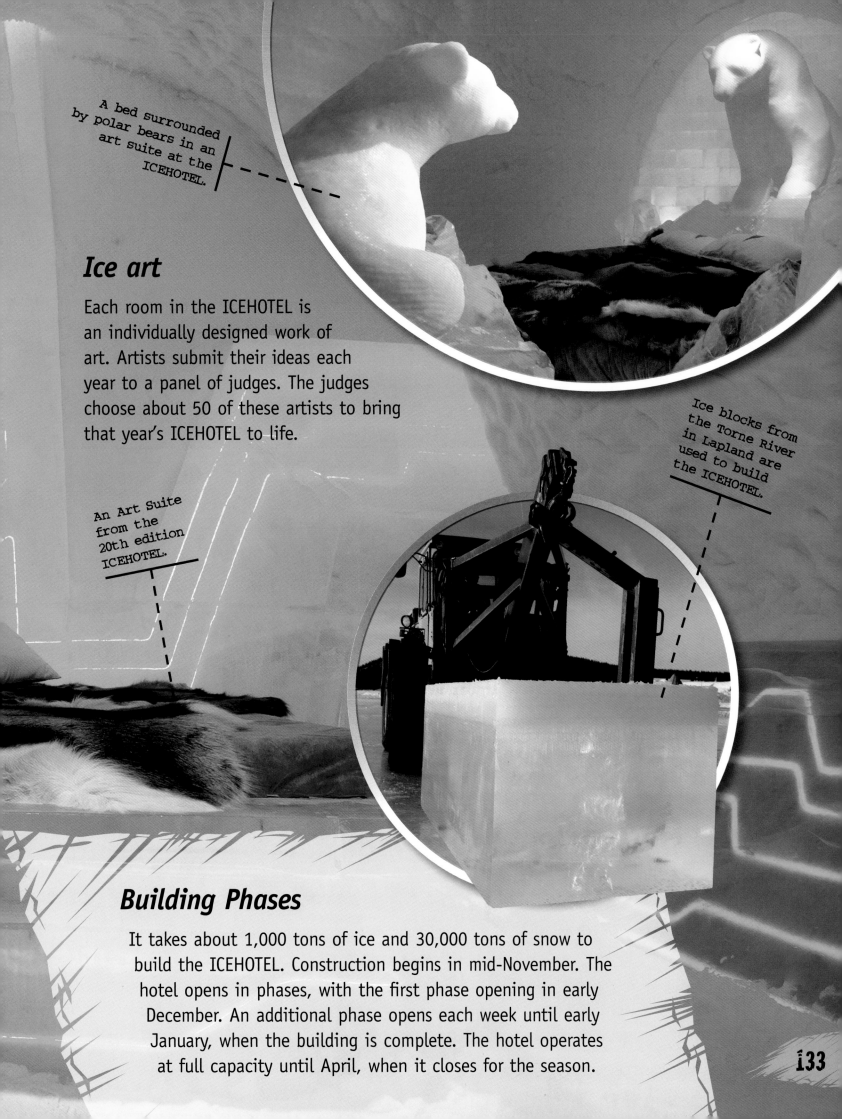

A bed surrounded by polar bears in an art suite at the ICEHOTEL.

Ice art

Each room in the ICEHOTEL is an individually designed work of art. Artists submit their ideas each year to a panel of judges. The judges choose about 50 of these artists to bring that year's ICEHOTEL to life.

An Art Suite from the 20th edition ICEHOTEL.

Ice blocks from the Torne River in Lapland are used to build the ICEHOTEL.

Building Phases

It takes about 1,000 tons of ice and 30,000 tons of snow to build the ICEHOTEL. Construction begins in mid-November. The hotel opens in phases, with the first phase opening in early December. An additional phase opens each week until early January, when the building is complete. The hotel operates at full capacity until April, when it closes for the season.

INTERNATIONAL SPACE STATION

Floating in orbit above Earth's surface is a satellite of pressurized living modules, laboratories, and work stations. Called the International Space Station (ISS), this structure is a permanent space base where astronauts study space science and try to find out more about what life in space does to our bodies. Astronauts live there for three to six months at a time.

The International Space Station on its orbit around Earth.

Astronauts on a spacewalk, working on the International Space Station.

AMAZING!

The ISS is the ninth space station to be inhabited. The first, Salyut, was launched in 1971.

Building Project

As well as performing their scientific work, the astronauts who go to the ISS build more parts onto it. Building began in 1998, and astronauts have been living there since 2000. Today, the station includes living quarters, science labs, and storage and service decks. It also has docking bays and an air lock so that visiting spacecraft can link to it, and astronauts can **spacewalk** outside.

Life in Space

The astronauts have to live with very low gravity, where things float around. Food and drinks are sucked from closed bags. The toilets have powerful suction to pull waste away. The crew members rest in sleeping bags that are tied to the walls.

135

BURJ KHALIFA

Burj Khalifa in Dubai is the world's tallest manmade structure. Completed in 2010, this incredible building rises 2,722 feet above ground level. It eclipses the previous world record holder, the KVLY-TV mast in North Dakota, U.S., which is 2,063 feet from bottom to top.

Burj Khalifa is the tallest building in the world.

AMAZING!

Twenty-two million hours of labor went into the construction phase of Burj Khalifa.

The Burj Khalifa cladding system is designed to withstand Dubai's intense summer heat.

Point of Pride

Burj Khalifa is the brainchild of Sheikh Mohammed bin Rashid Al Maktoum, the Prime Minister of the United Arab Emirates. Al Maktoum ordered architects to design "the greatest neighborhood known to man" with Burj Khalifa as its mighty centerpiece. He states that the building is a point of pride not only for the Emirates, but for the entire Arab world.

The view of downtown Dubai from the Burj Khalifa.

Highest Jump

At Burj Khalifa on April 21, 2014, Vince Reffet and Fred Fugen broke the Guinness World Record for the highest **BASE jump** from a building. The men leaped from a specially built platform at 2,717 feet, just five feet shy of the building's **pinnacle**.

GLOSSARY

Abolition
The legal ending of a practice or system.

Acidic
Containing acid. An acid is a chemical substance that gives off positively charged hydrogen ions when dissolved in water.

Acqua alta
An unusually high tide that periodically floods the Venetian lagoon. It occurs when the moon's pull combines with strong winds to push extra water into the region.

Acrophobia
A terror of heights. Most people have some fear of heights.

Artifact
A human-made object, usually old, that has historic or archaeological significance.

Automated
Able to operate independently, usually with the help of machines, without human supervision.

Bacteria
Microscopic, single-celled creatures that are neither plants nor animals. They were probably the first life forms to exist on Earth.

BASE jump
To parachute from a fixed structure or a cliff. BASE is an acronym that describes four common jumping platforms: buildings, antennae, spans, and Earth (cliffs).

Bioluminescence
The natural light emitted by certain creatures, such as fireflies and deep-sea fishes.

Caldera
A wide, flat, bowl-shaped crater left behind after a volcanic eruption.

Calving
A process in which ice chunks break off of a glacier and fall into a body of water.

Catacombs
Human-made caves originally built for religious ceremonies.

Chemosynthesis
A process by which living organisms use energy from chemical reactions to produce food, often in the absence of sunlight.

Colony
A group of bats that lives, breeds, and migrates together.

Crevasse
A deep crack in a glacier or an ice sheet. A crack in rock is called a crevice.

Cryptid
An animal whose existence has been suggested by the public, but has not been verified by the scientific community.

Debris
Scattered fragments, typically of something wrecked or destroyed.

Deck
The roadway or the pedestrian walkway surface of a bridge.

Deforestation
The cutting down and clearing away of trees from forests.

Deity
A god, goddess, or other divine being.

Dormant
Temporarily inactive. A dormant seed will not sprout until it encounters the right set of conditions.

Dune
A dune is a hill made of sand. Desert dunes are created by wind and coastal dunes by water.

Ecosystem
A community of animals and plants sharing an environment with non-living things such as water, soil, or climate, so that both animals and environment benefit from the relationship.

Emissions
Polluting substances discharged into the air by the technologies used in transportation, manufacturing, and other human activities.

Eroding
Wearing away through natural processes of erosion.

Erosion
The process of wearing away over time by wind, water, and other natural forces.

Evaporate
To change from a liquid to a gaseous form.

Evolution
A theory that states that living species change and diversify over many generations in response to natural conditions and pressures.

Evolve
To change and diversify over many generations in response to natural conditions and pressures.

Fjord
A long, narrow, deep inlet of the sea between high cliffs, usually formed by glacial melting.

Fumarole
An opening in or near a volcano that emits hot, sulfurous gases.

Geoglyph
A pattern or picture created on the ground by manipulating elements of the landscape.

Glacier
A massive block of ice that is moving very slowly.

Gorge
A steep valley between two cliffs, usually created by the flow of water over millions of years

Habitat
The natural home or environment of a plant, animal, or other living organism.

Humidity
A measure of the amount of water vapor in the air at a given time and place.

Hydrothermal vent
Deep under the sea, a chimney-like structure that spews out mineral-rich hot water that supports various life forms.

Iceberg
A chunk of ice that has broken off from a glacier. Icebergs are much larger below sea level.

GLOSSARY

Island gigantism
A biological phenomenon that occurs on islands. Island animals sometimes grow much larger than their mainland relatives due to lack of competition and predators.

Karst topography
A type of landscape formed when limestone and some other types of rocks dissolve, forming underground sinkholes and caves.

Larvae
Worm-like, immature forms of certain insects, such as gnats.

Lava
Magma that has erupted into the open air from a volcano.

Lintel
A horizontal support that spans and carries the load above an opening.

Magma
Molten (hot and liquefied) rock below Earth's surface.

Mausoleum
A building that holds a tomb or tombs.

Migrate
To move from one place to another and then back again on a regular schedule, often yearly, usually for reasons of food or climate.

Mineral
A mineral is a natural solid made of a particular mix of chemicals.

Moai
Huge rock statues probably depicting ancestors on Easter Island, Polynesia.

Mummies
A mummy is a body that has been preserved from decay, either accidentally or deliberately.

Naturalist
A person who studies nature and natural history, especially relating to plants or animals.

Notorious
Famous or well-known, typically for a bad quality or deed.

Organism
An individual animal, plant, or single-celled organism; a living being.

Penal
Relating to punishment, penalties, or prison institutions.

Pharaoh
The common title of the kings of ancient Egypt.

Pinnacle
The highest point.

Polygonal
Having the features of a polygon, which is a closed plane shape bounded by three or more line segments.

Polyp
A tiny, coral-building creature.

Promontory
A high ridge of land or rock that juts out into a body of water.

Radiotelegraphy
The transmission of information via radio waves.

Rusticle
Stalactites of rust, made of dissolving metal, that hang from the Titanic's wrecked hull.

Sarsen
A type of large sandstone boulder commonly found in certain parts of the United Kingdom and used in the construction of Stonehenge.

Sedimentary rock
Rocks that form when material is deposited on Earth's surface or within bodies of water, then cemented into a solid mass over time.

Sinkhole
A depression or hole in the ground caused by a collapse of the surface layer. This forms after the underlying rock wears away due to chemical processes, running water, or both.

Spacewalk
An astronaut moving around in space outside the spacecraft is on a spacewalk. May also be called extravehicular activity (EVA).

Stalactite
A deposit of minerals that hangs from the ceilings and walls of limestone caves. Deposits that grow upward from the ave floor are called stalagmites.

Summit
The highest point of a hill or mountain.

Tepui
A table-top mountain or mesa found in certain South American highlands.

Terraces
Flat steps built into the side of a mountain or hill, often used for agriculture.

Totora
A reed that grows in shallow parts of Lake Titicaca and other lakes of the Andes Mountains.

Vent
An opening that allows air, gas, or liquid to pass through.

Visible spectrum
The portion of the electomagnetic spectrum that is visible to the human eye. Rays in this range are called visible light or simply, light.

Vortex
A mass of spinning air, liquid, or dust that forms a hole at its center.

Wetland
A region where the soil is saturated with water. Marshes and swamps are common wetland features.

INDEX